The Mines and Geology of the Loomis Quadrangle Okanogan County, Washington

Washington Geological Survey

with an introduction by Kerby Jackson

This work contains material that was originally published in 1972 by the Washington Geological Survey.

Introduction

It has been over forty years since the Washington Geological Survey released it's important publication "Geology and Mineral Deposits of the Loomis Quadrangle". First released in 1972, this important volume has been out of print and has been unavailable to the mining community since those days, with the exception of expensive original collector's copies and poorly produced digital editions.

It has often been said that "*gold is where you find it*", but even beginning prospectors understand that their chances for finding something of value in the earth or in the streams of the Golden West are dramatically increased by going back to those places where gold and other minerals were once mined by our forerunners. Despite this, much of the contemporary information on local mining history that is currently available is mostly a result of mere local folklore and persistent rumors of major strikes, the details and facts of which, have long been distorted. Long gone are the old timers and with them, the days of first hand knowledge of the mines of the area and how they operated. Also long gone are most of their notes, their assay reports, their mine maps and personal scrapbooks, along with most of the surveys and reports that were performed for them by private and government geologists. Even published books such as this one are often retired to the local landfill or backyard burn pile by the descendents of those old timers and disappear at an alarming rate. Despite the fact that we live in the so-called "Information Age" where information is supposedly only the push of a button on a keyboard away, true insight into mining properties remains illusive and hard to come by, even to those of us who seek out this sort of information as if our lives depend upon it. Without this type of information readily available to the average independent miner, there is little hope that our metal mining industry will ever recover.

This important volume and others like it, are being presented in their entirety again, in the hope that the average prospector will no longer stumble through the overgrown hills and the tailing strewn creeks without being well informed enough to have a chance to succeed at his ventures.

Kerby Jackson
Josephine County, Oregon
January 2015

CONTENTS

Page

Abstract 1

Introduction 3

    Fieldwork 3

    Previous work and acknowledgments 3

Regional geologic setting 4

    Metamorphic rocks 4

    Plutonic rocks 4

    Tertiary sedimentary and igneous rocks 5

    Tectonic setting 5

    Sketch of local geologic history 6

Stratigraphy of the metamorphic rocks 6

    General features 6

    Order of superposition 6

    Nomenclature 7

    Redefinition of Anarchist Series and Kobau Group 8

    Anarchist Group 8

        Spectacle Formation 8

        Bullfrog Mountain Formation 10

        Sedimentation 11

    Palmer Mountain Greenstone 11

        Petrography 18

        Origin and age 18

    Mafic intrusive rocks 19

        Petrography 19

Specific gravity and chemical composition of the mafic intrusive rocks and the

    Palmer Mountain Greenstone 20

Kobau Formation 20

    Lithology 21

    Basal contact and thickness 22

    Age 22

CONTENTS

Stratigraphy of the metamorphic rocks—Continued   Page

Ellemeham Formation 22

    Lithology and stratigraphy 23

    Metamorphism 23

    Age 24

Plutonic rocks 25

    Serpentinite 43

    Whisky Mountain pluton 43

    Loomis pluton 44

        Contacts 45

        Compositional zoning 46

        Petrochemistry 46

        Age 46

    Anderson Creek pluton 47

    Toats Coulee pluton 49

        Contacts and border phase 49

        Petrochemistry 51

        Age 51

    Similkameen composite pluton 52

        Quartz monzonite—granodiorite 53

        Pyroxenite, malignite, and syenite gneiss 54

        Greisen, aplite, and alaskite 54

        Petrographic trends 55

        Petrochemistry 56

        Origin and age 56

    Petrologic summary of the plutonic rocks 56

Shankers Bend diatreme 58

Intrusive rocks in dikes and small masses 59

Rocks of Tertiary age 59

    Sedimentary rocks 59

        Conglomerate and arkose 60

        Volcanic graywacke 60

CONTENTS

Rocks of Tertiary age—Continued   Page

    Sedimentary rocks—Continued

        Age 61

           Interpretation of Tertiary sedimentation 62

    Igneous rocks 62

        Lithology 62

        Age 62

Metamorphism 63

    Anarchist Group and Kobau Formation 63

    Ellemeham Formation 64

Structural geology 65

    Major folds 66

        Wannacut Lake area 66

        Chopaka and Grandview Mountain areas 66

        Cayuse Mountain area 67

    Minor folds and crenulations 69

    Pre-Tertiary faults and lineaments 70

        Chopaka-Aeneas lineament 70

        Gold Hill fault 70

        Cayuse Mountain thrust fault 71

    Tertiary warping and faulting 72

    Interpretation of the structural record 72

Physiography and glaciation 72

    Chopaka Mountain landslide threat 76

Mineral deposits 77

    Metallic deposits 77

        Vein deposits 77

        Magmatic deposits 120

        Placer deposits 120

    Nonmetallic deposits 120

        Limestone 120

        Gypsite 121

References cited 122

CONTENTS

ILLUSTRATIONS

Plate   1. Geologic map and sections of the Loomis quadrangle .... In pocket

        2. Map showing mineral deposits of the Loomis quadrangle, Washington.................. In pocket

Page

Figure 1. Index map of Washington showing location of the Loomis quadrangle........................ 3

        2. Stratigraphic nomenclature of rocks of the Loomis quadrangle 7

        3. Specimen of distinctive chert-bearing conglomerate common to both the
Spectacle and Bullfrog Mountain Formations 9

        4. Amygdaloidal Palmer Mountain Greenstone from near the eastern contact
of the formation on Palmer Mountain 11

        5. Graphs showing specific gravities of specimens of Palmer Mountain Green-
stone and mafic intrusive rocks 12

        6. Coarsely recrystallized Palmer Mountain Greenstone from the body on Chopaka Mountain....... 12

        7. Silica variation diagrams of mafic samples of Palmer Mountain Greenstone and
mafic intrusive rocks 13

        8. Silica variation diagram showing the "alkali-lime index" for the least
metamorphosed samples of the suite of analyzed rocks from the Palmer
Mountain Greenstone and mafic intrusive rocks 21

        9. Classification scheme and modal plots of quartz-rich granitic rocks from the
Loomis quadrangle 25

     10. Map of the Loomis quadrangle showing pluton outlines, sample localities, and
contoured specific gravity data from the Loomis and Similkameen plutons 26

     11. Map of the Loomis quadrangle showing pluton outlines, sample localities, and
contoured quartz content of the Loomis and Similkameen plutons 27

     12. Map of the Loomis quadrangle showing pluton outlines, sample localities, and
contoured mafic mineral content of the Loomis and Similkameen plutons 28

     13. Map of the Loomis quadrangle showing pluton outlines, sample localities, and
K-feldspar as a percentage of total feldspar in the Loomis and Similkameen plutons 29

     14. Granodiorite of the Loomis pluton 45

     15. Plots of normative quartz, orthoclase, and albite plus anorthite of quartz-bearing
granitic rocks of the Loomis quadrangle 46

     16. Silica variation diagram comparing rocks from the Loomis and Toats Coulee plutons
and the Shankers Bend diatreme, with the Similkameen composite pluton and
some average rocks of Nockolds (1954) 47

     17. Silica variation diagram comparing the alkali-lime indices of the Similkameen,
Toats Coulee, and Loomis plutons 48

     18. Granodiorite of the Toats Coulee pluton 50

     19. Porphyritic quartz monzonite of the Toats Coulee pluton 50

     20. Silica variation diagram of the Similkameen composite pluton........................ 51

CONTENTS

ILLUSTRATIONS—Continued

Page

21. Granodiorite of the Similkameen composite pluton 53

22. Specimen of trachytoid malignite from dike cutting pyroxenite.......................... 55

23. Specimen of shonkinite showing typical xenomorphic-granular texture.................. 55

24. Normative trends of three plutons in the Loomis quadrangle compared with
    normative trends of some batholiths in the western United States 57

25. Fenitized greenstone breccia from Shankers Bend diatreme 58

26. Map showing main structural elements within the Loomis quadrangle and
    contiguous areas 65

27. Small-scale chevron folds in metagraywacke of the Spectacle Formation 67

28. Asymmetrical minor folds in limestone of the Spectacle Formation 68

29. Lower hemisphere equal-area stereographic projection of axes of 91 minor
    folds and crenulations in the Loomis quadrangle and contiguous areas 69

30. U-shaped valley of the Similkameen River, looking northward from the
    east flank of Chopaka Mountain... 74

31. View of Spectacle Lake, looking westward from southwest of gauging station 75

32. Looking west at sparsely forested crest of Chopaka Mountain 76

TABLES

Table   1. Petrographic summary of metamorphic rocks of the Loomis quadrangle .... In pocket

         2. Chemical and spectrographic analyses, modes, and norms of the Palmer
            Mountain Greenstone and mafic intrusive rocks.................................. 14

         3. Chemical and spectrographic analyses, norms, and modes of the Loomis
            and Toats Coulee plutons 30

         4. Chemical and spectrographic analyses, norms, and modes of the
            Similkameen composite pluton.. 34

         5. Chemical and spectrographic analyses, norms, and modes of rocks from the
            Shankers Bend diatreme and greenstone from the Ellemeham Formation 40

         6. Sequence of structural events in the Loomis quadrangle 73

         7. Mineral deposits of the Loomis quadrangle 78

GEOLOGY AND MINERAL DEPOSITS

of the

LOOMIS QUADRANGLE,

OKANOGAN COUNTY, WASHINGTON

By C. Dean Rinehart and Kenneth F. Fox, Jr.

ABSTRACT

The Loomis quadrangle is in north-central Washington at the eastern margin of the Cascade Mountains. It is underlain in its western half by plutonic rocks and in its eastern half by a clastic eugeosynclinal sequence named the Anarchist Series by Daly (1912).

Daly's Anarchist Series is redefined herein as the Anarchist Group, Palmer Mountain Greenstone, and Kobau Formation. The Anarchist Group comprises two formations, the Spectacle Formation and the superjacent Bullfrog Mountain Formation, and totals 20,000 feet in thickness. The Spectacle Formation consists of slate, metasiltstone, metasandstone, metalimestone,

and metaconglomerate, and contains a Permian, possibly Late Permian, fauna. The Bullfrog Mountain Formation is lithologically similar to the Spectacle except for a virtual absence of metalimestone. Unconformably overlying the Anarchist Group are the Palmer Mountain Greenstone and the Kobau Formation. They total 9,000 feet in thickness, are composed of greenstone, metasiltstone, and metachert, and are of Permian or Triassic age. These rocks were metamorphosed to the greenschist facies, probably in the mid-Triassic, and were intruded by quartz diorite in the Late Triassic and by granodiorite during the Early Jurassic.

The Ellemeham Formation of Jurassic or Cretaceous age overlies all the older metamorphic rocks on a pronounced angular unconformity. It is composed of siltstone, greenstone, and penecontemporaneously derived conglomerate, aggregating 3,000 feet in thickness. The Ellemeham was very weakly metamorphosed during a second metamorphic episode.

One of the three major plutons in the quadrangle, the Similkameen composite pluton, is unique because it is bordered by alkalic rocks—syenite and malignite— that grade concentrically inward through quartz-free and quartz-poor rocks, ultimately into normal granodiorite. As herein defined, the Similkameen includes Daly's "Kruger alkaline body" and part of his "Similkameen batholith" (1906, 1912). Although this composite pluton has not been satisfactorily dated as yet, a small alkalic diatreme, probably satellitic to the pluton, cuts the Ellemeham Formation. The emplacement of the diatreme was accompanied by intensive brecciation and fenitization of the country rock.

Locally fossiliferous conglomerate and arkose of early Eocene age, 4,000 feet thick, cover a few square miles in the northeastern part of the quadrangle where they unconformably overlie metamorphic rocks. They are intruded by dacite plugs that yield radiometric ages of about 50 million years.

Deformation of the metamorphic rocks that are older than the Ellemeham has produced fairly broad arcuate folds, which trend northwest, and upon which are superimposed northeast-trending folds and minor folds and crenulations. The minor folds and crenulations show a perceptible concentration of north to northeast trends. Faults and lineaments are numerous and of diverse type and age. A major north-northwest-trending lineament flanks the older plutonic rocks on the east and, along with an older pluton, is offset by a northeast-trending strike-slip fault. A north-trending,

west-dipping thrust fault extends most of the length of the quadrangle and apparently is younger than the major folds. Normal faults are ubiquitous but most show only small offsets and are traceable over only short distances.

Relief in the quadrangle is moderate on the east; it increases westward, attaining a maximum of more than 6,000 feet locally along the west boundary. The landscape owes much of its present form to glaciation during the Pleistocene. Drift remnants cover large areas in the central and eastern parts of the quadrangle; boulders, cobbles, and local patches of gravel indicate that even the highest summits were glaciated. The dominant topographic feature is a steep-walled, U-shaped valley that extends north-south through the quadrangle and strikingly marks the eastern front of the Cascade Mountains. Sinlahekin Creek flows through the southern part of this channel at the present time. Two large east-trending valleys intersect the master channel; they appear to have resulted from diversion caused by temporary damming of the main channel by ice or drift. The northernmost valley captured the Similkameen River and is its present channel.

The igneous and metamorphic rocks of the quadrangle are host to numerous small deposits of metallic minerals; most of these have been prospected and a few were mined around 1890. Gold, silver, and copper were the chief metals produced, mostly from quartz veins of the fissure-fill type, although a massive sulfide copper deposit was mined with some success during the early 1900's. Attitudes of the veins make no recognizable pattern in the quadrangle as a whole; in smaller areas, however, there is some consistency. The mineralization probably took place between mid-Cretaceous and early Tertiary time. At present, the areas that appear to be most favorable for prospecting are the Gold Hill-Deer Creek and Palmer Mountain areas.

INTRODUCTION

The Loomis quadrangle encompasses about 200 square miles in north-central Washington and adjoins British Columbia on the north (fig. 1). The western

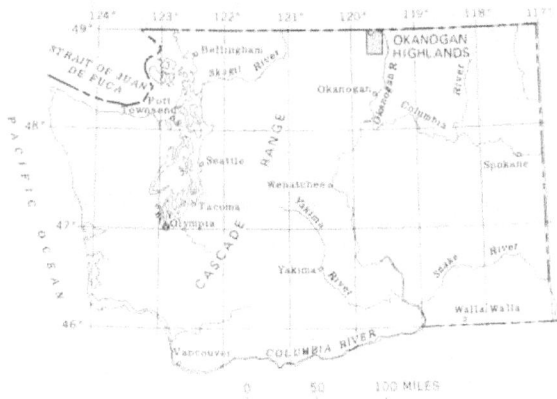

FIGURE 1.—Index map of Washington showing location of the Loomis quadrangle (shaded).

fourth of the quadrangle is occupied by the steep forested slopes of the Okanogan Range, the eastern-most of the ranges that make up the composite Cascade Range. The eastern three-fourths of the quadrangle is occupied by sparsely forested uplands that form the western margin of the Okanogan Highlands, a broad area of similar uplands that extends several tens of miles to the east and southeast. The Similkameen River is the only through-going drainage; it flows south from British Columbia, bends abruptly east, and joins the Okanogan River at Oroville, a few miles east of the quadrangle. Habitation is sparse and is about equally distributed among scattered ranches in the principal valleys and the small villages of Loomis and Nighthawk. Cattle grazing and apple growing are the mainstays of the local economy, though summer tourists attracted to the area by its scenic beauty and recreational possibilities also contribute significantly.

FIELDWORK

Study of the Loomis quadrangle was begun in 1963, as part of a cooperative program between the Washington State Division of Mines and Geology and the U.S. Geological Survey, in order to provide integrated knowledge of the geology and mineral deposits in a formerly active and moderately productive mining area in northern Okanogan County. Fieldwork began in July 1963 and was completed in August 1965. T. L. Fyock assisted for 4 weeks in 1963. Mapping was done largely on aerial photographs and compiled, by inspection, at a scale of 1:48,000.

PREVIOUS WORK AND ACKNOWLEDGMENTS

Previous workers have studied varied aspects of the local geology in various parts of the quadrangle over a long period of time, and we are considerably in their debt. During the first summer, R. W. Adams (1962) generously made available to us his geologic map, which covers about 30 square miles in the southeastern part of the quadrangle, and personally guided us to critical outcrops. Discussions in the field with him and with M. J. Hibbard (1962) proved most helpful in familiarizing us with the local geology. The graduate dissertations of R. W. Lounsbury (1951) and H. A. Pelton (1957) were also consulted throughout our study.

Because most of the mines are inaccessible, data bearing on the economic geology that had been recorded and compiled by earlier workers proved invaluable, particularly the publications of Hodges (1897), McIntyre (1907), Umpleby (1911b), Handy (1916?), and Huntting (1955, 1956). Additional unpublished information on mines and prospects was made available to us through access to the files of the Washington State Division of Mines and Geology.

The classic works of Smith and Calkins (1904), Daly (1912), and Waters and Krauskopf (1941) served as the basis for relating the geology of the quadrangle to that of the region. In addition, a geologic map of the adjacent Keremeos quadrangle in British Columbia by H. S. Bostock (1940) was helpful in correlating units and nomenclature with those used by Canadian geologists.

Since about three-fourths of the quadrangle is privately owned, fieldwork would have been difficult or impossible without the cooperation of the local ranchers. We express our appreciation to those who allowed us access to otherwise posted land.

The manuscript benefited considerably from painstaking reviews by R. C. Greene, H. C. Wagner, and R. G. Yates, and we gratefully acknowledge their efforts.

REGIONAL GEOLOGIC SETTING

METAMORPHIC ROCKS

The Loomis quadrangle straddles part of a belt of crumpled and folded metamorphic rocks crowded between the Colville batholith on the east and the multiple intrusive plutons of the Okanogan Range on the west (pl. 1, inset, in pocket). The metamorphic rocks of the area encompassed by this regional geologic map are of three general types: (1) eugeosynclinal deposits metamorphosed to the greenschist or amphibolite facies but retaining their original lithologic character; (2) gneisses and schists, probably largely higher metamorphic grade equivalents of the metamorphosed eugeosynclinal deposits; and (3) paragneisses and orthogneisses similar to those of the Shuswap terrane (Jones, 1959).

The weakly metamorphosed eugeosynclinal deposits include rocks of both Permian and Permian or Triassic age. The Permian rocks are mainly those of Daly's (1912) Anarchist Series as redefined in this bulletin. The Anarchist is a thick sequence of interbedded and interfingering weakly metamorphosed chert-conglomerate, graywacke, siltstone, and limestone within which fossils of Permian age have been found; it is regarded as a correlative of the Cache Creek Series of central British Columbia. The rocks of Permian or Triassic age are believed to be correlative with rocks known to unconformably overlie the Anarchist elsewhere. They may be correlative also with rocks not associated with the Anarchist, from which Triassic fossils have been found, such as the Nicola Group near Hedley, British Columbia (Rice, 1947) and the "lime belt" near Riverside, Washington (Waters and Krauskopf, 1941). The Permian or Triassic rocks constitute a thick and varied sequence

of weakly metamorphosed lavas, pyroclastics, tuffaceous sediments, massive cherts, and, locally, carbonate rocks. The carbonate rocks are a unique sequence that occurs in abundance only at Hedley and Riverside, but at these localities they are the chief lithologic types. If truly correlative, the Hedley and Riverside rocks constitute a limy marine facies of an otherwise mainly volcanic succession.

The "highly metamorphosed" rock shown on plate 1 (inset) includes paragneiss, schist, amphibolite, and marble that are known to grade into Anarchist rocks southwest of the Loomis quadrangle. The gradation to gneiss is marked by both increase in metamorphic grade and increased evidence of deformation through a zone several miles wide. Elsewhere it is likely that the high-grade metamorphics include rocks of other lineage.

Rocks of the Shuswap terrane are gneisses typified by regular, rather flat-lying foliation and conspicuous and equally regular lineation over areas of many square miles. The type area of the Shuswap (originally called "Shuswap Series" by Dawson, 1890) is near Shuswap Lake, about 130 miles north of the Loomis quadrangle. The Shuswap has been studied over much of its extent by Jones (1959), who suggested the correlation of the gneiss shown on plate 1 (inset) with the Shuswap to the north. White (1959, p. 69) has indirectly suggested that the rocks of the Colville batholith are also Shuswap, since they show similar lithologic and structural features.

PLUTONIC ROCKS

Most of the plutons are of batholithic dimensions and typically exhibit sharp crosscutting contacts.

The plutonic rocks belong chiefly to the quartz monzonite-granodiorite-quartz diorite clan. The Similkameen composite pluton is noteworthy because it grades in composition from quartz-rich rocks in its interior to alkalic rocks rich in mafic minerals at its borders. Similar alkalic rocks occur in small plugs or stocklike bodies that, along with the larger zoned plutons, define west-northwest-trending belts. The northeast corner of the Loomis quadrangle overlaps one such belt, and the alkalic rocks peculiar to it are more fully described in the body of this report. A parallel belt of similar mafic alkalic rocks lies 10 to 12 miles north-northeast at Oliver, British Columbia.

   Colville batholith.— The Colville batholith is a large granitoid mass containing, in the central part, gneissose porphyritic granodiorite that grades to non-foliated nonporphyritic rock in the eastern part of the region and to layered paragneiss along its western contact. The foliation and layering typically undulate from horizontal to dips of about 15°, and the rock typically exhibits a pervasive west-northwest lineation. While the origin of the mass is somewhat controversial, Waters and Krauskopf (1941) showed that it was force-fully emplaced as a semisolid mass and Snook (1965) demonstrated that the layered gneisses are of metamorphic origin.

   The plutonic rocks are of Triassic(?) to Cretaceous age, except possibly for one small pluton 9 miles east of Skaha Lake. Little (1961) correlated this pluton with the Coryell batholith of Tertiary age, which lies to the east. The oldest dated pluton discovered within the region is the Loomis pluton of Late Triassic age (this paper).

TERTIARY SEDIMENTARY AND IGNEOUS ROCKS

   Conglomerates, arkoses, graywackes, and tuffs, capped by or interbedded with andesitic to dacitic lavas, unconformably overlie the crystalline basement. In the Loomis quadrangle and at other localities, the sedimentary deposits are cut by small stocks similar in composition to the lavas. K-Ar dating of correlative igneous rocks of southern British Columbia by Mathews (1964) indicates that they crystallized during a single rather short period of the Eocene, although rocks of a younger episode of intrusion are present to the north in central British Columbia. Recent dating by Obradovich (this paper) of comparable rocks of the Loomis quadrangle establishes that they are virtual age equivalents of those dated by Mathews.

TECTONIC SETTING

   The region illustrated on plate 1 (inset) includes part of the central portion of the Mesozoic orogen, a north-northwest trending belt of igneous intrusives and folded sedimentary and metamorphic rocks, which is approximately bounded on the east by the Purcell Trench of northern Idaho and on the west by the chain of Cenozoic volcanoes and intrusive rocks of the Cascades. Current knowledge of the tectonic history of northeastern Washington and northern Idaho has been summarized by Yates and others (1966), that of northwestern Washington by Misch (1966), and that of British Columbia by White (1959). Therefore, only a few points particularly relevant to the structural geology of the Loomis quadrangle are considered here.

   The Permian and Triassic(?) eugeosynclinal sediments of the central part of the orogen, which includes the Loomis area, underwent strong deformation, metamorphism, and magmatism during orogeny in the Triassic Period (White, 1959). The age of the initial orogenic deformation and metamorphism of the Paleozoic rocks of the western flank has not been closely bracketed, but "the minimum age . . . appears to be Late Permian (-Early Triassic?)" (Misch, 1966, p. 102). Faults, fold axes, and axes of elongate plutons of the Triassic orogeny and similar linear elements of possibly older orogenies strike predominantly northwest, imparting a pronounced northwest grain to the map. Initial deformation of orogenic proportions on the eastern flank of the orogen may have been somewhat later, since the folds, faults, and axes of plutons strike predominantly northeast, and were considered by Yates and others (1966, p. 52) to be of Jurassic and Cretaceous age.

   The plutonism that began during the Triassic, in the central part of the orogen at least, continued intermittently without a well-defined maximum into the Tertiary. Misch (1966) recognized a period of orogenic deformation of Cretaceous age in the Cascades that produced extensive overthrusts and other faults of great length and displacement; structural trends are to the north-northwest. The Methow-Pasayten region, whose eastern edge is shown in the southwest corner of the inset on plate 1, marks the site of Cretaceous marine deposition and concomitant faulting, again showing north-northwest trends (Misch, 1966). Misch suggests that the source of some of the Cretaceous clastics of the Methow-Pasayten region lies to the east. If so, the central part of the orogen has probably been an area of positive relief since Late Cretaceous time and, except for mild warping and north to north-northeast faulting of Eocene or younger age, has been little deformed since then.

SKETCH OF LOCAL GEOLOGIC HISTORY

The geologic history that can be reconstructed in the Loomis quadrangle begins in the Permian with the deposition of marine epiclastic rocks of the Anarchist Group, which comprises the Spectacle and Bullfrog Mountain Formations, in a rapidly subsiding geosynclinal trough. Deposition was halted by uplift of the Anarchist terrane, and subsequent erosion locally removed much of the Bullfrog Mountain Formation. The erosion surface was buried under the lavas, pyroclastics, and bedded cherts of the Palmer Mountain and Kobau Formations, which accumulated in a eugeosynclinal marine environment during the Permian or Triassic.

Orogenesis here, as farther north in British Columbia, probably began during the Middle or Late Triassic and ended the depositional phase. The entire sequence was tightly folded, metamorphosed to the greenschist facies, and, during the Late Triassic, intruded by the Loomis pluton; during the Early Jurassic it was intruded by the Toats Coulee pluton. The earlier folds were subsequently refolded during a second main period of folding before epeirogenic uplift and erosion brought the orogenic cycle to a close.

Erosion beveled the plutonic and folded metamorphic terrane, and on this surface the lava and pyroclastic rock of the Ellemeham Formation were deposited.

The Similkameen composite pluton intruded the sequence in Jurassic or Cretaceous time, thermally metamorphosing along its contact the rocks of the Kobau and Ellemeham Formations. At about the same time, the Anarchist Group to the south was somewhat telescoped by a north-trending, west-dipping thrust that traverses about two-thirds the length of the quadrangle. Many of the metallic mineral deposits were formed at about this time. Subsequent erosion deroofed the Similkameen pluton by Eocene time. Proof of this comes from the northeastern part of the quadrangle

where cobbles of the pluton were recognized in basal beds of a clastic sequence intruded by plugs of dacite of Eocene age. Deformation followed or coincided with the sedimentation and associated volcanism during the Tertiary, resulting in mild eastward tilting of the clastic beds and some normal faulting. The last major geological event—one which essentially constitutes the final chapter in the local geologic history— was extensive glaciation during the Pleistocene and the consequent sculpture of the landscape and establishment of the drainage system that exists today.

STRATIGRAPHY OF THE METAMORPHIC ROCKS

GENERAL FEATURES

Metamorphic rocks in the Loomis quadrangle have been only weakly metamorphosed, so that their original compositional and textural character has not been entirely erased. They originally were a thick sequence of siltstones, pyroclastics, and lavas with interlayered massive chert that overlay a still thicker sequence of lenticular interfingering and interbedded conglomerate, graywacke, siltstone, limestone, and slate. Table 1 (in pocket) presents a petrographic summary of the metamorphic rocks.

Although they are thick locally, individual lithologic units seldom have great lateral continuity. Except for the limestones, which crop out as distinctive light-gray cliffs and ledgy slopes, the metamorphics are rather monotonous, drab greenish-gray and brown rocks. These features, coupled with the blurring of lithologic contrasts as a consequence of metamorphism, have hindered the precise definition of the stratigraphic sequence and have thus impaired the solution of structural complexities.

ORDER OF SUPERPOSITION

The establishment of the order of superposition in the Loomis quadrangle, fundamental to an accurate stratigraphic and structural interpretation, is limited by scanty and often conflicting evidence. The order of superposition given here is based primarily on such depositional features as channeling and crossbedding, which are found only rarely and at widely scattered localities and exclusively within the Bullfrog Mountain Formation. The channels are generally small notches with widths from one to several inches cut at contacts between coarse and fine clastics. Crossbedding is rare, and generally poorly defined. A few features resembling flame structures were found in thinly laminated siltstones. Graded bedding was noted at a few localities, but was often discredited by conflicting indications at the same locality. The inference that beds were overturned at one locality was corroborated by bedding-cleavage relations in a laminated slate and metasiltstone sequence.

NOMENCLATURE

The oldest rocks in the area are correlative with rocks called the Anarchist Series by Daly (1912, p. 389) for the Anarchist mountain plateau, located 8 miles northeast of the Loomis quadrangle. According to Daly, "The name is literally not inappropriate, for these rocks cannot as yet be reduced to stratigraphic order or structural system." Other workers were able to clarify the stratigraphic succession, and that given here owes much to the efforts of our predecessors, particularly Umpleby (1911b) and Waters and Krauskopf (1941).

Although the Anarchist as originally defined included virtually all of the metamorphic rocks in the vicinity, the original scope of the name has since been reduced. Bostock (1940) named rocks in Canada along

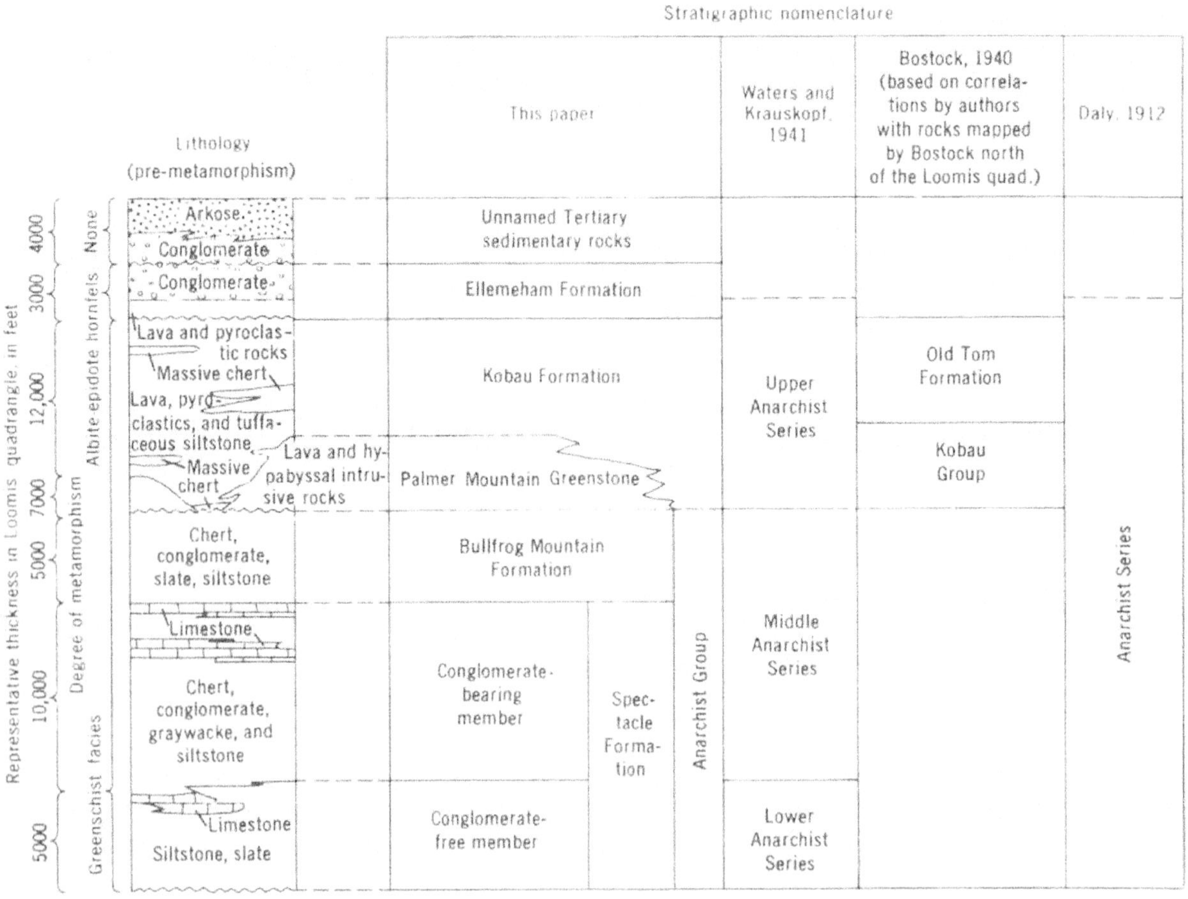

FIGURE 2.—Stratigraphic nomenclature of rocks of the Loomis quadrangle.

the International Boundary the Kobau Group for Mount Kobau, 8 miles north of the Loomis quadrangle. On Daly's (1912) maps these rocks lie within his Anarchist Series. Little (1961) subsequently honored Bostock's Kobau, but retained the term Anarchist Group for rocks that were east of the area mapped by Bostock. Waters and Krauskopf (1941) divided and mapped Daly's Anarchist as the Lower, Middle, and Upper Anarchist Series. The lower part of their Upper Anarchist Series is contiguous with Bostock's Kobau Group (fig. 2).

REDEFINITION OF ANARCHIST SERIES AND KOBAU GROUP

Our mapping in the Loomis quadrangle has revealed a substantial unconformity between the metavolcanic rocks and quartzites of Waters and Krauskopf's Upper Anarchist Series and the subjacent metamorphosed epiclastic rocks. The regional significance of the unconformity is unclear, since only the rocks below are accurately dated. Rather than include with the Anarchist lithologically distinctive rocks above the unconformity, it seems advisable to restrict the Anarchist to the beds below the unconformity and adopt Bostock's Kobau for the beds above. Therefore, Bostock's name Kobau is herein adopted, but is redefined as the Kobau Formation and includes only the rocks above the unconformity; the lower part of Daly's Anarchist Series is here designated the Anarchist Group in accordance with modern usage.

ANARCHIST GROUP

Spectacle Formation

The Anarchist Group is divided into the Spectacle and Bullfrog Mountain Formations, names herein assigned to mappable units within the Anarchist. The Spectacle Formation takes its name from Spectacle Lake where part of the formation is exposed. The type locality is the area roughly bounded by Wannacut Lake on the east and Palmer Mountain on the west (pl. 1). The thickness is at least 15,000 feet, based on calcula-

tions from the map and cross sections (pl. 1) that make allowance for known faults and structural conditions. The base is not exposed. The formation is conformably overlain by the Bullfrog Mountain Formation.

The bulk of the Spectacle Formation consists of metamorphosed interbedded sharpstone conglomerate, limestone, graywacke, siltstone, and black slate. The conglomerate is restricted to the upper 10,000 feet of the unit and is nearly ubiquitous within this interval. On the basis of this conglomerate, the Spectacle is subdivided into an upper conglomerate-bearing member and lower conglomerate-free member. The contact between the two members probably represents a facies change because field relations suggest both lateral and vertical gradation and interfingering of the conglomerate with siltstone and graywacke. Marked divergence of the contact with an underlying limestone bed was noted in secs. 26 and 36, T. 39 N., R. 26 E.

Conglomerate-free member. — Rocks of the conglomerate-free member of the Spectacle Formation exhibit the dynamic effects of regional metamorphism to a somewhat greater extent than those of the overlying formations. Crenulated and crumbly pale-green phyllites and similarly folded black slates predominate and are interbedded with less deformed massive to thin-laminated pale-green graywackes and gray- to pale-green siltstones. Cleavage or schistosity considerably obscures or obliterates the primary fabric. Where bedding can be detected, it is usually paralleled by the schistosity, but exceptions are numerous. Although in the Loomis quadrangle the conglomerate-free member is mainly a sequence of fine metaclastics, correlative rocks to the east (Oroville quadrangle) and south (Conconully quadrangle) contain interbeds of greenstones thought to be metavolcanic rocks.

We infer that the rocks of the Spectacle Formation represent an accumulation of fine clastics derived from unknown highlands, presumably deposited in a shallow marine environment. The presence of greenstones suggests that the scene was occasionally punctuated by outpouring of lava and pyroclastic material.

Conglomerate-bearing member. — The metaconglomerates of the conglomerate-bearing member attest to the onset of more vigorous sedimentation. The clasts of the conglomerates are mostly sharply angular to subrounded dark-gray chert that are firmly bonded by a light-gray matrix of coarse lithic graywacke (fig. 3). The chert clasts thus contrast strongly with the matrix, which results in a rather striking texture, and is an exception to the otherwise rather featureless metamorphic rocks of the Anarchist Group.

0 ____ 3 cm

FIGURE 3.—Specimen of distinctive chert-bearing conglomerate, a lithologic type common to both the Spectacle and Bullfrog Mountain Formations. Darkest fragments are chert; others are mostly varied fine-grained epiclastic rocks.

Despite their widespread distribution and stratigraphic importance, the conglomerates constitute only a small proportion of the upper part of the Spectacle Formation. The bulk of the unit consists of rocks similar to those of the conglomerate-free member; that is, metamorphosed interbedded slate, siltstone, graywacke, and limestone.

The limestones are resistant rocks and stand in relief above the more easily eroded adjacent metasedimentary rocks. In addition, their light-gray tone contrasts sharply with the other rocks of the Anarchist, permitting easy recognition from a distance or on aerial photos. Despite the extreme irregularity of any particular limestone unit, they are as a group confined to the conglomerate-bearing member.

The limestone units are lenticular or podlike, commonly ranging from 100 to 150 feet in thickness,

but locally are as much as 2,500 feet thick; along the strike, the units pinch and swell, bifurcate, and either wedge out or grade into calcareous graywacke or siltstone. The limestones are composed chiefly of calcite and are locally fairly pure, although more commonly they are cherty, sandy, shaly, or graphitic. Bedding is commonly well developed and is usually thinly laminated and regular, but at a few places it is scoured and exhibits weak crossbedding. The bedding is locally crumpled or folded, and in a few places is isoclinally folded; typically it is conformable with the enclosing strata. Except for scattered crinoid columnals, the limestones are very fine grained to fine grained; prior to metamorphism they would probably have been classed as micrite (Folk, 1959). Presumably the limestones accumulated in sites of low hydraulic energy, perhaps as lime muds in shallow, quiet waters. The upper limit of the limestone beds defines the upper contact of the Spectacle Formation.

Fossils.— Associated with the limestone in the uppermost beds of the Spectacle Formation is a thick, very massive, gray- to grayish-green, fine- to coarse-grained, calcareous metagraywacke from which distorted but recognizable fossils have been obtained at several localities. The fossils are brachiopods, pelecypods, and gastropods and occur as molds resembling oyster and (or) snail shells, and are found by overturning talus blocks and examining the lichen-free underside of the rock.

Collections from localities shown on the map were submitted to J. T. Dutro, Jr., and E. L. Yochelson, U.S. Geological Survey. They reported that "several of the brachiopods [from USGS locality No. 21912-PC, NW$\frac{1}{4}$NW$\frac{1}{4}$ sec. 23, T. 38 N., R. 26 E.] are sufficiently well preserved to permit a definite age assignment. One excellent external mold of the brachial valve of a productoid is identified as Megousia. Another large mold is most probably a Yakovlevia (Muirwoodia of earlier usage)." They concluded that the age of the collection was Permian, possibly early Late Permian. Fossils from other localities were too poorly preserved to allow precise identification, but in general included fossils that resembled Permian types (J. T. Dutro, Jr., and E. L. Yochelson,

written communication, May 18, 1965). These localities are USGS Nos. 21911-PC, SW¼SE¼ sec. 4, T.
39 N., R. 26 E.; and 21913-PC, SE¼SW¼ sec. 27,
T. 39 N., R. 26 E. The age of the Spectacle Formation is therefore considered to be Permian, perhaps
Late Permian.

Dutro and Yochelson's findings confirm Girty's
earlier assignment of lithologically similar beds in the
adjacent Oroville quadrangle to the Permian (In Waters
and Krauskopf, 1941, p. 1364), and establish the general age equivalence of the Anarchist Group (as used
herein) to the Permian Cache Creek Group (Armstrong,
1949, p. 47-51; Thompson and others, 1950, p. 52-
53) of British Columbia. This correlation was suggested
by Smith and Calkins (1904, p. 22) and by Daly (1912,
p. 559) on the basis of lithologic similarities.

Bullfrog Mountain Formation

The Bullfrog Mountain Formation is composed of
rocks similar to those of the Spectacle Formation, except that the Bullfrog has no limestone. Interbedded
slate, metasiltstone, metagraywacke, and sharpstone
metaconglomerate make up the bulk of the unit. The
formation is named for Bullfrog Mountain, herein designated the type locality, where the formation is well
exposed, and its stratigraphic position between the
conformably underlying Spectacle Formation and the
unconformably overlying Kobau Formation and Palmer
Mountain Greenstone is easily demonstrated. The
Bullfrog Mountain Formation is about 5,000 feet thick
at the type locality.

The rocks of the Bullfrog Mountain Formation
are metaclastic, with grain sizes ranging from clay
size in the slates, to boulder size in the metaconglomerates; slate and metasiltstone are the most abundant
rocks. The slates are typically black to dark-gray,
rather incompetent rocks, and are poorly exposed. The
siltstones are hard and dense, dark gray to pale green,
and thinly laminated to massive. Locally the bedding
in the metasiltstones is crumpled, convoluted, slumped
or erratic due to penecontemporaneous deformation.

The Bullfrog Mountain Formation contains a conspicuous, though minor, proportion of metagraywacke

and metaconglomerate. These are usually hard, resistant rocks, and hence are well exposed. They are
typically medium gray to pale green, massive, and
interbedded with slate or siltstone. The metaconglomerates are composed of angular fragments of diverse
size and type, bonded by a graywacke matrix, and
similar in appearance to the specimen illustrated in
figure 3. Dark-gray chert clasts are the most numerous, but slate, siltstone, and volcanic rock fragments
are also abundant. Limestone pebbles are rare but several can usually be found in each outcrop. The clasts
attain a maximum size of 1 foot in length. Single conglomerate beds up to 200 feet thick are present, but
the usual thickness of individual beds is in the 1- to
5-foot range. The conglomerates grade both laterally
and vertically into finer clastics.

The character of the Bullfrog Mountain Formation
changes somewhat toward the southwestern part of the
quadrangle. Thick sequences of dark-gray to black
massive vitreous quartzite and interbedded thin-bedded
to thinly laminated quartzite and thinly laminated
greenstone predominate, with subordinate metaconglomerate and metagraywacke. Because the formation appears to thicken to the southwest, it is likely that these
beds represent portions of the upper part of the Bullfrog
Mountain that were removed by pre-Kobau erosion from
the section at the type locality at Bullfrog Mountain.
Unfortunately, the crosscutting Loomis pluton separates
these rocks from the rest of the formation to the north,
and primary lithologic details are difficult to compare
because the southern rocks are somewhat masked by
greater metamorphic recrystallization. For example,
andalusite and sericite, pseudomorphous after andalusite, were found at several places on Aeneas and
Douglas Mountains.

The formation is about 5,000 feet thick at the
type locality but thins abruptly to the northeast and
apparently thickens to the southwest. The variation in
thickness may be at least partly a depositional feature,
but presumably some variation is due to erosion that
occurred prior to deposition of the overlying beds of
the Kobau. The unconformity beneath the Kobau has
not been observed in outcrop, but field relations north
of Bullfrog Mountain suggest considerable discordance.
Because of its general lithologic similarity and grada-

tional contact relations with the Spectacle Formation,
the Bullfrog Mountain Formation is also considered to
be Late(?) Permian in age.

Sedimentation

Individual beds within the Anarchist Group do
not have great lateral extent; instead, the overall pattern suggests a sequence of interfingering and overlapping lenses of epiclastic rock and subordinate lenses of
limestone and volcanic rock. The overall thickness is
substantial, probably in excess of 15,000 feet.

Constituents of the clastic rocks are poorly sorted
and subangular, and some are rather coarse, indicating
a short distance of transport and a nearby source. These
characteristics, together with the occurrence of marine
fossils, suggest rapid accumulation under marine conditions, necessarily in a continually subsiding basin.
Presumably the sources of the sediment were nearby
highlands experiencing orogenic uplift. The metalavas
of the lower part of the Anarchist Group and the small
but ubiquitous proportion of clasts of volcanic origin
indicate that both the highlands and the depositional
basin experienced some concomitant volcanism.

Moderate uplift of the basin brought the depositional phase of the Anarchist Group to a close and
resulted in development of an erosional surface on the
uplifted strata. Subsequent resumption of subsidence
was accompanied by the deposition of siltstones, pyroclastics, lavas, and massive cherts of the Kobau Formation and volcanic rocks of the Palmer Mountain Greenstone, which are lithologic types markedly different
than those in the underlying Anarchist.

PALMER MOUNTAIN GREENSTONE

The Palmer Mountain Greenstone, here named
for the mountain on which it is well exposed, consists
of a metamorphosed assemblage of mafic igneous rocks,
generally basaltic in composition, that show wide variations in texture. Very fine to fine-grained, medium- to
dark-greenish-gray greenstone (fig. 4), amphibolite,
metadiabase, and felsic variants of the metadiabase,

1 cm

FIGURE 4.—Amygdaloidal Palmer Mountain Greenstone from near the eastern contact of the
formation on Palmer Mountain; rock probably
originated as a lava flow or a shallow intrusion.

are the most common rocks, and presumably represent
metavolcanic rocks—including pyroclastic rocks—and
their metamorphosed intrusive equivalents. Greenstone
and metadiabase are distinguished from each other on
the basis of grain size—greenstone is aphanitic, metadiabase is coarse enough to allow recognition of felty
or diabasic texture with the aid of a hand lens. The
rocks are massive and generally hard, dense, and resistant to erosion, except along shear zones where they
are weakened by closely spaced fracture cleavage and
local mylonitization. Schistosity is generally absent.
The bluffs near the common corner of secs. 16, 17, 20,
and 21, T. 39 N., R. 26 E., are herein designated the
type locality, because good exposures of the varied
types can be observed there. Assemblages of lithologically similar rocks on Bullfrog Mountain and Chopaka
Mountain are here correlated with and assigned to the
Palmer Mountain Greenstone.

At Palmer Mountain, the Palmer Mountain Greenstone is at least 7,000 feet thick if the basal contact
extends to depth at dips approximating those of the
underlying Anarchist Group and the contact connects
across beneath Palmer Mountain in the manner shown
in cross section C-C', plate 1. If the formation is more

4848484848848484848

tightly folded than shown on the cross section, the thickness would be greater. The abrupt thinning of the underlying Bullfrog Mountain Formation north of Spectacle Lake strongly suggests that the basal contact of the Palmer Mountain Greenstone is unconformable. The contact is not exposed over much of its length but field relations show that it grossly parallels bedding in the Bullfrog Mountain Formation, though with some irregularity. A small angular discordance was noted on the northeastern flank of Aeneas Mountain. Bedding in the Palmer Mountain Greenstone was noted at only a few localities near the south and east margins of the unit, generally close to the basal contact. There, the bedding is represented by wispy thin laminae contrasting slightly in color or texture with adjacent strata. The

bedded zones are interlayered with massive aphanitic greenstone, suggesting interbedded pyroclastics and flows.

From fine-grained greenstone along the north, east, and south margins, the grain size progressively increases toward the center and western margin of the mass where medium-grained metagabbro is common. The increase is in part gradational and in part due to an increase in abundance of coarse-grained varieties. Gradation from fine- to medium- and even coarse-grain size can be seen in many exposures, but sharp contacts between rocks of contrasting grain sizes are more common; in the latter, age relations are typically equivocal although evidence of both coarse intruding fine and the reverse were observed. Where intrusive relations are exposed, the younger rocks are typically more leucocratic than the host, although the gabbroic center and western margin of the mass on Palmer Mountain is a notable exception.

The range in composition is indicated by the range in specific gravities (fig. 5). The specific gravities of several specimens from any given locality typically vary within such wide limits that even gross trends within the unit could not be established with certainty.

The Palmer Mountain Greenstone on Chopaka Mountain consists mostly of amphibolite that is char-

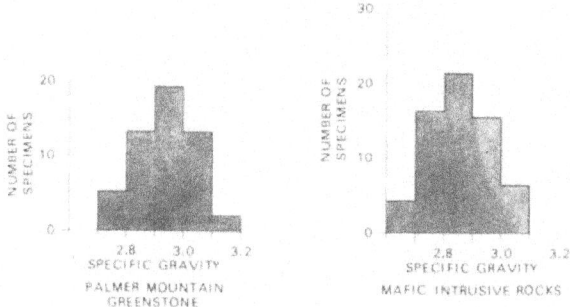

FIGURE 5.—Graphs showing specific gravities of specimens of Palmer Mountain Greenstone and mafic intrusive rocks.

FIGURE 6.—Coarsely recrystallized Palmer Mountain Greenstone from the body on Chopaka Mountain. Rock is an amphibolite showing evidence of extensive recrystallization and mobilization.

acterized by irregular layers or bands of contrasting grain size or color index (fig. 6). Its relations to the adjacent serpentinite and Kobau Formation are mostly equivocal, but at one locality it appears to intrude the Kobau. On Chopaka Mountain the rocks are more extensively recrystallized than they are on Palmer Mountain and primary features have been largely obliterated. The layers generally are gradational in detail and are commonly folded, offset, curled, or swirled. They are generally between 1 and 3 inches thick, and they thin and thicken or grade to a different texture along strike. The finest grained rocks are aphanitic, the coarsest pegmatitic. The amphibole weathers in relief, resulting in a pale-green to dark-green rock studded with stubby to oval black crystals. Massive, coarse-grained amphibolite without layering is locally abundant.

On Chopaka Mountain the unit is cut everywhere by an intricate network of ramifying and anastomosing veinlets of pinkish-white to greenish-white or green zoisite, feldspar, and calcite. The veinlets typically do not have sharp walls, but grade through thicknesses of about half a centimeter into the host rock. In addition, they share local structural offsets and irregularities with the other elements of the fabric, suggesting that they developed concomitantly with the general deformation and metamorphism. Such an extensive network of partly gradational veinlets makes it seem unlikely that the metamorphism was isochemical. Nevertheless, a chemical analysis of an amphibolite from one of the areas of apparently less altered rock (L-402-C, table 2 and fig. 7) shows considerable similarity to an analyzed specimen of metagabbro (L-319-A) from Palmer Mountain.

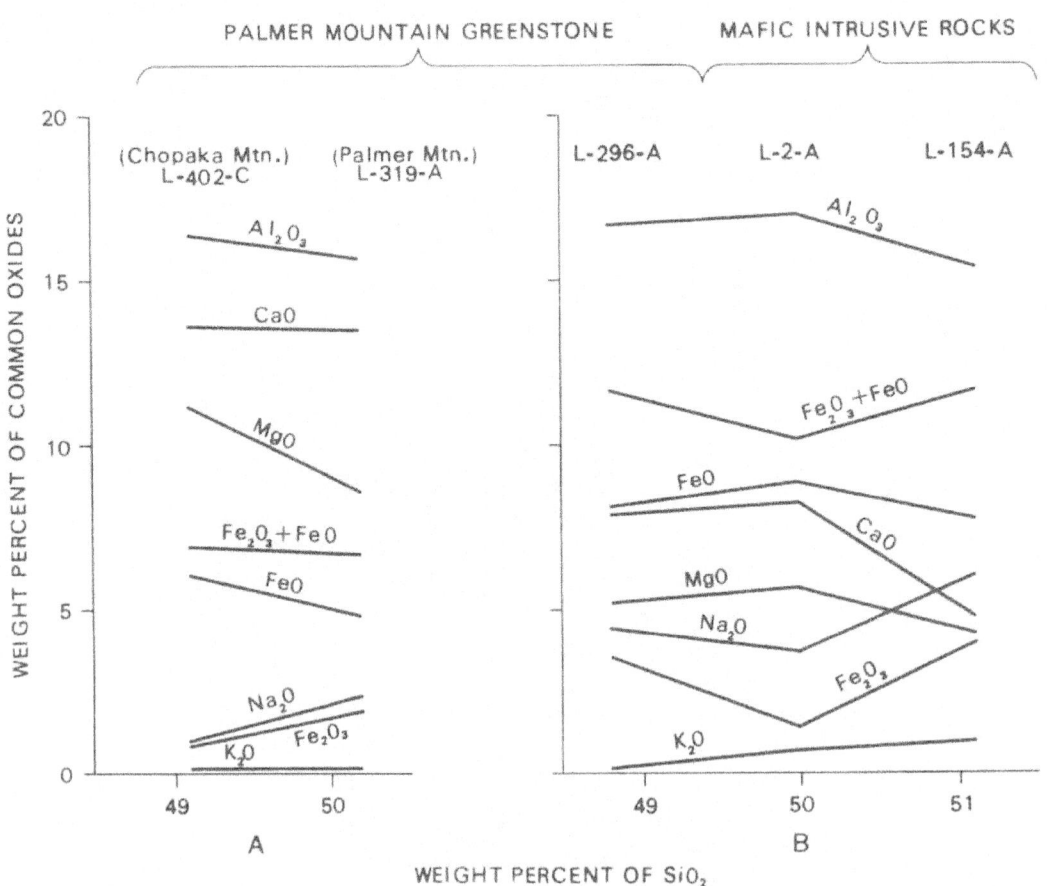

FIGURE 7.—Silica variation diagrams of mafic samples of Palmer Mountain Greenstone and mafic intrusive rocks. A, samples of highly metamorphosed parts of Palmer Mountain Greenstone; B, samples of weakly metamorphosed parts of Palmer Mountain Greenstone and mafic intrusive rocks.

TABLE 2.—Chemical and spectrographic analyses, modes, and norms of the Palmer Mountain Greenstone and mafic intrusive rocks

	Palmer Mountain Greenstone				Mafic intrusive rocks		
	L-402-C[1] Amphibolite	L-319-A Metagabbro	L-296-A Metadiabase	L-192 Meta-quartz monzonite	L-17-A Meta-quartz monzonite	L-2-A Porphyritic greenstone	L-154-A Metadiabase
	Chemical analyses [2]						
SiO_2	49.1	50.2	48.8	66.4	61.4	50.0	51.1
Al_2O_3	16.4	15.7	16.7	15.7	17.3	17.0	15.4
Fe_2O_3	0.89	1.9	3.5	1.4	1.5	1.4	3.9
FeO	6.1	4.8	8.1	1.2	2.1	8.8	7.7
MgO	11.2	8.6	5.2	0.94	0.91	5.6	4.2
CaO	13.6	13.5	7.8	2.4	3.6	8.2	4.7
Na_2O	0.99	2.4	4.4	5.3	4.0	3.7	5.1
K_2O	0.10	0.11	0.13	3.1	2.9	0.58	0.90
H_2O^-	0.13	0.10	0.26	0.09	0.12	0.18	0.21
H_2O^+	1.2	1.2	2.9	1.1	1.8	3.0	2.7
TiO_2	0.03	0.85	1.3	0.18	1.4	1.1	1.8
P_2O_5	0.00	0.06	0.22	0.08	0.18	0.17	0.22
MnO	0.17	0.15	0.23	0.08	0.14	0.19	0.16
CO_2	<0.05	0.25	0.10	1.3	2.5	0.05	1.2
Total	99.96	99.82	99.64	99.27	98.85	99.97	99.29
Specific gravity (lump)	2.92	2.94	2.82	2.55	2.65	2.87	2.79
Specific gravity (powder)	3.00	3.01	2.98	2.71	2.77	2.98	2.87

Norms (weight percent)

Quartz	...	...	...	20.7	25.1	...	1.8
Corundum	...	...	...	2.5	7.3	...	0.8
Orthoclase	0.6	0.6	0.8	18.5	17.2	3.4	5.4
Albite	8.4	20.4	37.4	45.2	34.0	31.3	43.5
Anorthite	40.0	31.8	25.5	3.2	0.9	28.1	14.4
Wollastonite	11.5	13.9	4.7	...	...	4.7	...
Enstatite	23.2	19.6	6.5	2.4	2.3	8.6	10.5
Ferrosilite	8.9	5.6	5.1	0.9	0.6	8.3	8.3
Forsterite	3.3	1.3	4.6	...	...	3.8	...
Fayalite	1.4	0.4	4.0	...	...	4.0	...
Magnetite	1.3	2.8	5.1	2.1	2.2	2.0	5.7
Ilmenite	2.8	1.6	2.5	0.3	2.7	2.1	3.4
Apatite	1.6	0.1	0.5	0.2	0.4	0.4	0.5
Calcite	0.1	0.6	0.2	3.0	5.7	0.1	2.8

Approximate mode

[M=major constituent; m=minor constituent; A, B, C, etc., = decreasing order of abundance]

Quartz	...	m	...	A	A	...	...
Microcline (microperthite)	...	...	...	A	...	...	...
Plagioclase Composition(An)	50 Bytownite(?)	50 Albite	M (15)	A (25-30)	B Albite	A Albite	50 Oligoclase
Biotite	...	...	...	D	...	...	1
Sericite	...	m	m	B	C	C	...
Actinolite	50 (incl. relict hbld.)	25	M	...	...	E	35
Augite	...	...	M	...	...	B	...
Chlorite	...	5	...	B	D	C	5
Epidote/ clinozoisite	...	15	M	m	...	...	2

(See footnotes at end of table.)

TABLE 2.—Chemical and spectrographic analyses, modes, and norms of the Palmer Mountain Greenstone and mafic intrusive rocks—Continued

| | Palmer Mountain Greenstone | | | | Mafic intrusive rocks | | |
	L-402-C[1] Amphibolite	L-319-A Metagabbro	L-296-A Metadiabase	L-192 Metaquartz monzonite	L-71-A Metaquartz monzonite	L-2-A Porphyritic greenstone	L-154-A Metadiabase
Approximate mode—Continued							
Zoisite	...	M	...	...	...	...	...
Calcite	...	m	m	C	E	...	m
Accessories	m	m	5	m	m	m	5
Semiquantitative spectroscopic analyses for minor elements [3] (C. Heropoulos, analyst)							
B	0	0	0	0	0	0.005	0
Ba	0.003	0.005	0.01	0.2	0.2	0.1	0.07
Co	0.005	0.003	0.003	0.0005	0.0007	0.005	0.005
Cr	0.05	0.07	0.02	0.002	0.0005	0.015	0.002
Cu	0.0007	0.007	0.007	0.0003	0.001	0.03	0.015
Ga	0.001	0.0015	0.002	0.0015	0.0015	0.0015	0.0015
La	0	0	0	0	0.003	0	0
Ni	0.015	0.015	0.007	0.001	0.003	0.005	0.002
Pb	0	0	0	0.003	0	0	0
Sc	0.007	0.01	0.007	0	0	0.007	0.005
Sr	0.01	0.03	0.03	0.3	0.1	0.07	0.05
V	0.015	0.03	0.02	0.007	0.005	0.05	0.05
Y	0	0.003	0.005	0.001	0.002	0.003	0.005
Yb	0	0.0003	0.0005	0.0001	0.0002	0.0003	0.0005
Zr	0	0.005	0.01	0.01	0.02	0.01	0.015

[M=major constituent, m=minor constituent; A, B, C, etc., =decreasing order of abundance]

1/ Sample locations:

L-402-C, NW$\frac{1}{4}$ sec. 30, T. 40 N., R. 25 E.

L-319-A, SE$\frac{1}{4}$ sec. 18, T. 39 N., R. 26 E.

L-296-A, SW$\frac{1}{4}$ sec. 28, T. 39 N., R. 26 E.

L-192, NW$\frac{1}{4}$ sec. 28, T. 40 N., R. 26 E.

L-71-A, SE$\frac{1}{4}$ sec. 14, T. 39 N., R. 26 E.

L-2-A, NW$\frac{1}{4}$ sec. 26, T. 38 N., R. 26 E.

L-154-A, SW$\frac{1}{4}$ sec. 3, T. 39 N., R. 26 E.

2/ Chemical analyses by P. Elmore, S. Botts, G. Chloe, J. Glenn, L. Artis, H. Smith, and D. Taylor, using the rapid method of Shapiro and Brannock (1956).

3/ Results are reported in percent to the nearest number in the series 1, 0.7, 0.5, 0.3, 0.2, 0.15, and 0.1, etc., which represent approximate midpoints of interval data on a geometric scale. The assigned interval for semiquantitative results will include the quantitative value about 30 percent of the time.

The Palmer Mountain Greenstone on Bullfrog Mountain is lithologically similar to rocks at the type locality on Palmer Mountain, although zoning was not recognized. Pyroclastic rocks were recognized locally in the southwestern part but the body is mainly composed of flows and intrusive rocks. Recrystallization is substantially less than at Chopaka Mountain, and relict textures similar to those at the type locality on Palmer Mountain are locally well preserved. On Bullfrog Mountain, however, relations of the Palmer Mountain Greenstone to rocks stratigraphically above and below are obscure.

Alaskitic to granodioritic dikes of uncertain affinity, a fraction of an inch to several feet thick, are common in all three bodies. Some contamination by greenstone, resulting in hybrid rock, was evident at several places. A few dikes show chilled borders. Intrusion of felsic rock has locally produced complex intrusion breccia; probably the most striking example is at the Black Bear mine on Palmer Mountain.

Petrography

The essential mineralogy of the entire unit is fairly constant, as most rocks are composed of varied proportions of hornblende/actinolite, albite/oligoclase, epidote/clinozoisite, chlorite, biotite, sericite, and calcite. Common accessories are sphene, apatite, ilmenite, and magnetite. A number of specimens contain sparse quartz and K-feldspar, apparently as essential minerals. Zoisite is a common mineral in the western part of Palmer Mountain and in the rocks on Chopaka Mountain, occurring both in veins and as a common product of plagioclase alteration in metagabbro. It was not found on Bullfrog Mountain. Relict clinopyroxene was found in the cores of several amphibole crystals, and the pyroxene-like shapes of amphibole crystals in several metagabbro specimens suggest pseudomorphs after pyroxene. Except for the relict pyroxene, the major minerals of the Palmer Mountain Greenstone are products of metamorphism and thus reflect metamorphic grade as well as the original composition of the rocks. The preponderance of amphibole, chlorite, and calcite and the scarcity

of potassium feldspar and quartz indicate that the original rock was mafic. The range in chemical composition is shown by the analyses in table 2. If the metamorphism has been essentially isochemical, the original rock that formed the bulk of the unit was probably andesite, though the unit included, as probable later differentiates, minor amounts of rock as silicic as quartz monzonite (specimen L-192).

Relict textures found in the less metamorphosed rocks range from fine-grained, locally porphyritic, trachytic, or felty texture in the greenstones to medium-grained diabasic texture in the metagabbro. The outline of former zoning in plagioclase phenocrysts is marked by fine-grained secondary minerals in some specimens.

Origin and Age

Except for minor pyroclastics, most of the Palmer Mountain Greenstone originated as flows and shallow intrusives. The proportions of each are uncertain because of recrystallization and the scarcity of primary features. The distribution of greenstone and diabase suggests that the Palmer Mountain Greenstone was a pile of mafic extrusive rocks—a volcano—into which mafic magma was abundantly intruded, cooling slowly enough in places to produce fairly coarse diabasic textures. This conclusion is in accord with the suggestion of Smith and Calkins (1904, p. 47) that the Palmer Mountain Greenstone is the exhumed core of an old volcano, and that this volcano may have been the source of much of the volcanic material found in the fringing sedimentary deposits.

The age of the Palmer Mountain Greenstone is unknown. Its relation to underlying rocks is clear only on Palmer Mountain where basal pyroclastic rocks are in depositional contact with the underlying Bullfrog Mountain Formation. Its lithology is similar to that of the Old Tom Formation as described briefly by Bostock (1940), who regards the Old Tom as Triassic or older but states that it lies stratigraphically above the Independence Formation where "fossils of doubtfully Mesozoic age have been found." In 1965, we reconnoitered exposures of the Old Tom that had been mapped by

Bostock near Keremeos, B. C., 15 miles north of the International Boundary, and we found greenstone and metadiabase similar to that exposed at Palmer Mountain. It seems reasonable, therefore, to tentatively regard the two as age correlatives; hence, the Palmer Mountain Greenstone is assigned an age of Permian or Triassic.

MAFIC INTRUSIVE ROCKS

The mafic intrusive rocks comprise numerous small elongate bodies, irregular in shape, whose compositions include greenstone, metadiabase, metagabbro, metadiorite, and less abundant felsic variants that are intruded into rocks of the Anarchist Group and Kobau Formation. They are, at least in part, the same as the metagabbro and metadolerite described by Waters and Krauskopf (1941). Most abundant is metadiabase which, in many places, grades into aphanitic greenstone or into the coarser grained metagranitoid rocks. Textures in the latter include diabasic, hypautomorphic-granular, and porphyritic. Patchy, erratic variation in grain size from fine to medium grained is a characteristic feature in nearly all bodies. Fragmental texture was seen in a very few places and in most of these it can be attributed to cataclasis. The rocks are generally massive, but locally are schistose where cut by shear zones or along contacts. Greenstone and metadiabase are typically greenish gray to dark greenish gray; other rocks of the unit range from light to dark gray. Pyrite is a fairly common and conspicuous accessory mineral in all rocks of the unit. Calcite is almost invariably present in amounts ranging from a trace to several percent. Quartz is rarely visible in hand specimen but in thin section is fairly common in small amounts.

The mafic intrusive rocks are elongate parallel to the strike of the bedded metamorphic units and most are concordant, with typically sharp contacts. Age relations are generally uncertain, but a few of the bodies have crosscutting contacts, some have gradational fine-grained borders that might be interpreted as chilled contacts, and many of the larger, grossly concordant bodies have apophyses and satellitic dikes that clearly intrude the metasedimentary rocks. Evidence of extrusive origin, such as flow breccias, pillow structure, and vesicular texture, is lacking. Thus an intrusive origin for most of the bodies seems most likely. Nevertheless, an extrusive origin for some of the greenstone cannot be ruled out completely, and indeed, the occurrence of fine-grained igneous fragments in conglomerates of the Anarchist Group indicate probable local volcanism before or during deposition of those units. Although rocks of different ages may have been inadvertently grouped together, the unit as a whole appears grossly equivalent to the Palmer Mountain Greenstone and the Kobau Formation, and is accordingly assigned a Permian or Triassic age.

Petrography

Primary igneous textures are generally recognizable, though extensively modified by metamorphism. Relict felty and diabasic texture was commonly detected in samples of the greenstone and metadiabase, but it was typically modified or destroyed by the later development of granoblastic texture. Large, subsequent relict feldspar phenocrysts (or saussurite pseudomorphs) were also recognized in many specimens. Thin mylonitic zones and mortar structure, noted in some thin sections, provide evidence of cataclastic deformation not apparent megascopically.

The mineralogic composition of the rocks is compatible with the greenschist metamorphic facies (treated under Metamorphism); the typical assemblage is: quartz, locally twinned albite, actinolite, calcite, chlorite, epidote/clinozoisite, sericite, and opaque minerals; common accessories are apatite, sphene, K-feldspar, and biotite. Relict clinopyroxene cores in amphibole crystals were noted in two specimens. In several other specimens, shapes of amphibole crystals, or cores of colorless amphibole surrounded with a green pleochroic variety, resemble pyroxene and are probably pseudomorphs. In several metagranitoid rocks former zoning of plagioclase is indicated by saussurite distribution within albite crystals.

SPECIFIC GRAVITY AND CHEMICAL COMPOSITION OF THE MAFIC INTRUSIVE ROCKS AND THE PALMER MOUNTAIN GREENSTONE

Both the mafic intrusive rocks and the Palmer Mountain Greenstone are compositionally heterogeneous, judging by the variation in color index from one body to the next and from outcrop to outcrop within a body. This variation is documented by measurements of specific gravity of a number of specimens, shown by histograms in figure 5. Comparison of the histograms shows that the variation within the mafic intrusive rocks is roughly comparable to that within the Palmer Mountain Greenstone, but that the average specific gravity of the Palmer Mountain Greenstone is somewhat greater than that of the mafic intrusives. No obvious trends were detected in the geographic variation of specific gravity.

Specimens that represent the range in specific gravity were selected for chemical analysis. The analyses (table 2) confirm that within broad limits the specific gravity is a function of chemical composition. Hence, the histograms of specific gravity qualitatively reflect the compositional variation of the rock.

The analyses include two specimens (L-402-C and L-319-A) representing metamorphic rocks that show great textural diversity; they are closest to contacts with granitic rocks, and are of somewhat higher metamorphic grade than the others. Inspection of table 2 and figure 7 reveals that the analyses of L-402-C and L-319-A are similar to each other but differ appreciably from analyses of the other rocks with comparable silica content. If these differences are attributed to differences in metamorphism, then the higher grade rocks have gained CaO and MgO and lost Na_2O and total iron, and exhibit a reduced ferrous-to-ferric iron ratio relative to the lower grade rocks. These compositional changes are qualitatively similar to those in amphibolites brought about by progressive metamorphism in the Adirondacks (Engel and Engel, 1962). Thus it is likely that the metamorphism of the higher grade parts of the Palmer Mountain Greenstone has not been isochemical and that the

diverse compositional types within it are in part products of metasomatism.

Assuming, however, that the less metamorphosed parts of the Palmer Mountain Greenstone (specimens L-296-A, L-192) represent a close approximation of the original composition, we infer (comparing analyses by inspection with the averages of Nockolds, 1954) that the bulk of the Palmer Mountain Greenstone was originally similar in composition to tholeiitic andesite with minor variants as silicic as dacite and perhaps as mafic as basalt. Similarly, the mafic intrusives resemble tholeiitic andesite and trachyandesite, with leucocratic variants as silicic as rhyodacite or dacite.

The general similarity of the Palmer Mountain Greenstone and the mafic intrusive rocks, their juxtaposition, and their probable rough equivalence in age suggest that the mafic intrusive rocks are hypabyssal intrusive equivalents of the Palmer Mountain Greenstone and thus are derived from a common magma. Speculating that this is correct, and viewing the group as a consanguineous suite, excluding the two most highly metamorphosed members, the diagram on figure 8 indicates that the suite is somewhat alkalic in character, with an alkali-lime index near the border of Peacock's (1931) alkalic and alkalic-calcic series.

KOBAU FORMATION

The name Kobau Group was first used by Bostock (1940) for quartzite, schist, and greenstone that underlie an area of more than 60 square miles immediately north of the Loomis quadrangle. It was named for Mount Kobau, a 6,100-foot mountain in British Columbia, 8 miles north of Lenton Flat. In the Loomis quadrangle the formation is about 12,000 feet thick and is well exposed on the northeast flank of Ellemeham Mountain in sec. 19, T. 40 N., R. 26 E., herein designated a reference locality. It is composed of a thick sequence of grayish-green phyllites, greenstones, and massive metacherts lying unconformably on rocks of the Anarchist Group, and probably in part conformably on the Palmer Mountain Greenstone. It is unconformably overlain by the Ellemeham Formation.

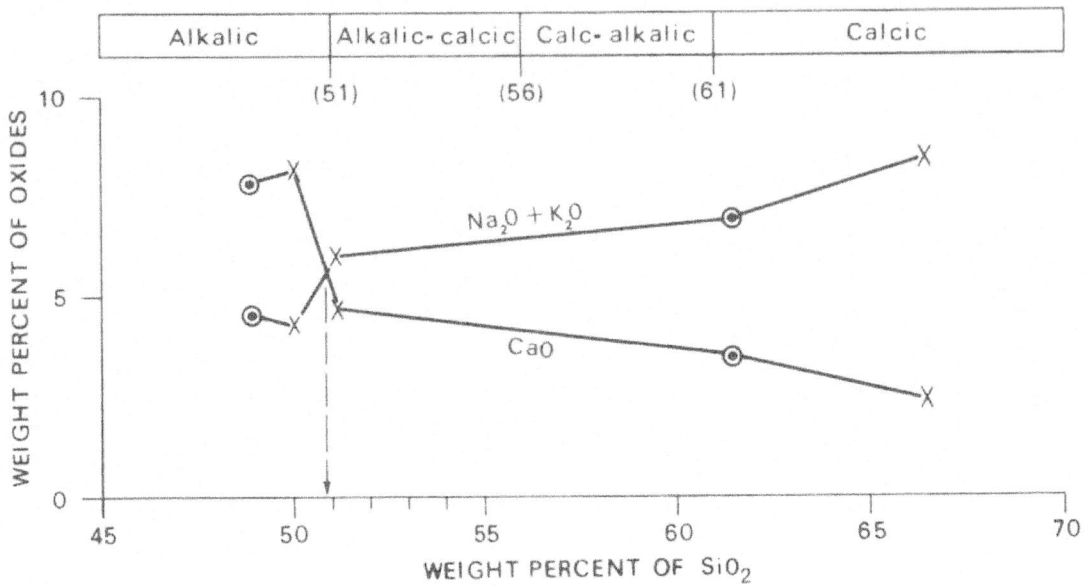

FIGURE 8.—Silica variation diagram showing the "alkali-lime index" (percent SiO_2 at which $Na_2O + K_2O = CaO$) for the least metamorphosed samples of the suite of analysed rocks from the Palmer Mountain Greenstone (⊙) and mafic intrusive rocks (x). (After Peacock, 1931.)

Lithology

Phyllite.— Although the relative proportions of phyllite, metachert, and greenstone vary considerably, phyllites are generally most abundant and metacherts and greenstones are subordinate but ubiquitous constituents. The phyllites appear to be weakly metamorphosed thin-bedded clastics, chiefly tuffaceous siltstones and fine-grained graywackes. Typical outcrops exhibit a sequence of irregularly alternating 5- to 20-foot-thick units that include grayish-green thin-laminated phyllitic siltstone, greenish-gray massive to thin-laminated, weakly phyllitic graywacke, and light-gray to grayish-green thin-bedded nonfoliated metachert, with scattered thin interbeds or partings of weak, crumbly phyllite. Except for the metacherts, individual units are themselves rather heterogeneous sequences marked by irregular finer scale alternations of thin beds of slightly different character.

Metachert.— Quartzites (metachert) are particularly prominent in the Kobau Formation. They range in color from opaque black to translucent bluish gray or light gray. They consist of fine-grained granoblastic quartz with a varied but minor proportion of

actinolite, mica, or other metamorphic minerals. Individual beds are massive to thin bedded, but on favorable exposures even the massive beds display relict traces of irregular thin bedding.

References to the association of massive cherts with lavas and greenstones are abundant in geologic literature, and by analogy it seems likely that the Kobau quartzites are metamorphosed cherts rather than metamorphosed sandstones. Relict textures attributable to a parent clastic rock such as sandstone are lacking. Because metamorphism has not succeeded in erasing the clastic texture of the siltstones and graywackes, the absence of clastic textures in the quartzites of the Kobau supports the hypothesis that the quartzites are derived from massive chert beds rather than sandstones.

The metacherts are present throughout most of the Kobau, usually in massive beds 5 to 100 feet thick but ranging down to thin intercalations an inch or less in thickness, interbedded with the other rocks. Transitional types were noted, ranging from rather impure greenish-gray metachert to grayish-green siliceous siltstones. The metacherts reach their extreme development on Ellemeham Mountain, where they crop out in beds up to 1,500 feet thick.

Greenstone. — The greenstones of the Kobau
are subordinate but nevertheless widespread. They are
generally interbedded with the phyllites and metacherts
but locally occupy considerable areas to the exclusion
of other rock types. They are typically thin-laminated
or massive grayish-green rocks that commonly contain
enough calcite to effervesce in dilute hydrochloric
acid. Microscopic examination revealed relict primary textures in a few specimens suggesting deriva-
tion from pyroclastic rocks, but in most specimens,
primary textures have been obliterated by metamorphism. It is likely, nevertheless, that the greenstones
include both metalavas and metapyroclastics.

Basal Contact and Thickness

The Kobau Formation appears to be at least
12,000 feet thick at the section exposed on Ellemeham
Mountain. This thickness represents a minimum figure,
since it was scaled from a cross section constructed
between the base of the formation in Longacre Draw
and the contact with the Similkameen composite pluton to the northwest. The Kobau probably in part
overlies the Palmer Mountain Greenstone along an
irregular contact, but it is also locally intruded by
it. Where the Palmer Mountain Greenstone is not
present, the Kobau rests unconformably on rocks of
the Anarchist Group. Furthermore, on Bullfrog Mountain the lower part of the Kobau appears to interfinger
with the upper part of the Palmer Mountain and thus
is probably in part its age equivalent. If this hypothesis is accurate, then the Kobau is best considered
as a belt of marine deposits that formed satellitic to,
and partly as a result of, the volcanic activity represented by the Palmer Mountain Greenstone.

Age

No fossils have been found in the Kobau Formation, but the fact that it unconformably overlies the
Anarchist Group of Late(?) Permian age indicates
that it is at least as young as Late Permian, and is
perhaps Triassic. Bostock (1940) shows the Kobau as

Carboniferous(?), and appears to base this age on the
fact that Permian fossils were found in the Blind Creek
Formation, the oldest limestone unit in a sequence that
he considered as overlying the Kobau. However, the
Blind Creek is in fault contact with the Kobau. In
view of the ambiguities, it seems appropriate to con-
sider the Kobau as Permian or Triassic.

ELLEMEHAM FORMATION

The Ellemeham Formation, here named for Ellemeham Draw in the northeast part of the quadrangle,
comprises a basal member of greenstone and tuffaceous
siltstone, an intermediate member of monolithologic
greenstone conglomerate, and an upper conglomeratic
member composed of heterogeneous fragments of metamorphic rock. All three units are well exposed about
2 miles south of Ellemeham Draw at the bluffs northeast
of Wannacut Lake (sec. 1, T. 39 N., R. 26 E.), herein designated the type locality. The unit is approximately 3,000 feet thick as measured on cross section
B-B' (pl. 1), but ranges widely in thickness, as do
the component members, throughout the area of exposure.

The lower members of the formation commonly
crop out as rounded, convex, irregular cliffs or bluffs
that form a line of irregular benches or gently north-
to east-sloping cuestas. The less resistant upper member forms poorly exposed irregular veneers on the
benches or is covered in concave slopes. From a distance the bluffs appear a rusty brown that contrasts
with the subdued gray or greenish gray of the subjacent Anarchist Group. At closer range the rocks of
the lower two members are olive to brownish gray on
weathered surfaces. Compact hand specimens are
obtained with difficulty because the rocks are weak
and brittle and they shatter along myriad fracture
planes at the blow of a hammer. This formation is,
in part, the same as a unit called "phyllite breccia"
by Lounsbury (1951), who remarked on its resemblance
to the basal Tertiary conglomerate. From thin-section
study, however, he recognized that, unlike the Tertiary, it had been metamorphosed and he therefore
included it with rocks of the Anarchist Group.

Lithology and Stratigraphy

The stratigraphically lowest member consists of massive greenstone and thinly laminated, locally calcareous siltstone and (or) tuffaceous siltstone. These types appear to intergrade and are also coarsely interbedded. At Shankers Bend in the northeastern part of the quadrangle, the unconformable contact with the underlying Kobau Formation is marked by a thin conglomerate consisting mostly of pebbles and cobbles of the Kobau; quartzite is the chief rock type among the clasts.

The massive greenstone comprises both tuffs and flows. Poorly defined bedding and local preservation of clastic texture suggest pyroclastic origin for some of the greenstone; elsewhere the presence of amygdules, suggestion of pillow structure, and fluidal texture indicate flows. In most of the greenstone, however, features such as these are too poorly preserved to permit identification of the original rock.

Bedded siltstone is much less common than massive greenstone but scattered outcrops are found throughout the lowest member. The siltstone contains thin calcareous laminae that, although nowhere abundant, are widely distributed. At a few localities northeast of Wannacut Lake near the east border of the quadrangle, thin lenses of limestone are interbedded with siltstone. Gradation to sandstone occurs at several places, and in some of these, calcite grains compose 10 to 15 percent of the rock. Quartz grains are typically present in moderate amounts. At Shankers Bend near the east portal of the railroad tunnel, layers of hornblende crystal tuff, which are a fraction of an inch to more than a foot thick, are interbedded in siltstone. In a few places these layers cut across bedding but are most common as thin, discontinuous layers parallel with bedding. On the whole, bedding is extremely sparse, and where observed shows gentle to moderate dips; the only folds seen are small scale and appear to be due to penecontemporaneous slump.

Progressing both laterally and stratigraphically upward in the formation, the siltstone and massive greenstone pass abruptly—though locally the transition occupies a few inches to a few feet—to coarse breccia of the monolithologic conglomerate member. Most of the breccia fragments range from sand to cobble size but locally are as large as a house. The breccia consists almost exclusively of fragments of the massive greenstone or siltstone of the lowest member set in a finer matrix of the same lithologic types. It grades upward to conglomerate as the fragments become somewhat rounded.

The youngest member of the Ellemeham is distinguished from the underlying monolithologic conglomerate member only because it contains numerous fragments of recognizable Anarchist and Kobau lithologic units. The contact is not sharp and typically is gradational over many tens or several hundreds of feet. Its location on the map marks the lowest stratigraphic occurrence of Anarchist and Kobau fragments. Anarchist and Kobau fragments progressively increase in abundance stratigraphically upward, but, except in the uppermost part, the major proportion of the rock appears to be derived from the greenstone and tuffaceous siltstone of the lowest member. Fragments of granitic rocks are absent.

The contact of the Ellemeham with the overlying rocks, a clastic sequence of Tertiary age, is not sharp. Its precise location is, in fact, somewhat arbitrary because locally the lowermost Tertiary beds are lithologically indistinguishable from the uppermost Ellemeham Formation, probably because the latter was reworked during earliest Tertiary sedimentation. Nevertheless our data, though admittedly scant, indicate that the Ellemeham Formation shows evidence of weak metamorphism, whereas rocks of Tertiary age show none. Unfortunately, we are presently able to make this distinction only by microscopic examination of thin sections.

Metamorphism

Although metamorphism of the Ellemeham Formation is weak, all of its members have been locally metamorphosed. Metamorphism ranges up to that of the greenschist facies, although in some areas the rocks reflect little more than diagenetic alteration. Rocks metamorphosed to the greenschist facies are composed of quartz, albite, epidote/clinozoisite, tremolite/actinolite, biotite, chlorite, ilmenite, and magnetite. Potassium feldspar and muscovite are less common constituents. The assemblage belongs to the lower part of the greenschist facies. Primary textures in the finest grained rocks have been obliterated by the metamorphic development of fine-grained grano-blastic textures. Metamorphic planar structures are restricted to the local growth of mica along bedding planes or to sporadic growth along planes of fracture cleavage. Relict hornblende phenocrysts in meta-volcanic rocks locally show metamorphic effects by patchy replacement of the hornblende by biotite, epidote, and tremolite or actinolite similar to grano-blastic amphibole in the matrix.

The following is a tentative interpretation of the depositional environment of the Ellemeham Formation:

(1) A basin, cut into Kobau and older rocks, of moderately deep water was receiving silt-size sediment together with voluminous amounts of fine volcanic ash, intermittently interrupted by lava flows. Local occurrence of limestone, scarcity of sand and absence of coarse detrital material, and lack of such features as crossbedding or ripple marks suggest the sediments accumulated in water of moderate depth.

(2) Late in this episode of quiet accumulation, these fairly thick, poorly lithified deposits became unstable, probably owing to nearby orogenic activity, and much of the upper part of the sequence slid to a lower part of the basin, undergoing fragmentation in the process. The disrupted, fragmented part of the sequence came to rest on both undisturbed Ellemeham sediments and folded Paleozoic metamorphic rocks. Subaqueous fragmentation and penecontemporaneous deformation seem to offer a better explanation for the distribution and monolithologic character of the frag-mental middle member, and the local convolute bed-ding and slump structures in the lowest member, than accumulation through ordinary subaerial processes.

(3) Shortly after or concurrent with this event, fragments of the older metamorphic rocks began to accumulate in the basin and became mixed with the upper part of the disrupted sequence. As deposition continued, the abundance of coarse metamorphic de-tritus, derived mostly from the underlying Kobau For-mation but also to a lesser degree from rocks of the Anarchist Group, progressively increased. No evi-dence was found to indicate whether this coarse unit—the uppermost member—was deposited subaqueously or subaerially, but the latter process could account for a progressive increase through time in abundance of Kobau and Anarchist fragments by normal downcutting of streams into the surrounding terrane. Termination of this episode was followed by erosion and subsequent deposition, chiefly of granitic detritus, in the Tertiary.

Age

No fossils have been found in the Ellemeham Formation, nor are there any rocks of known age near-by with which the Ellemeham can be lithologically correlated. Nevertheless, the following facts bear on its age:

(1) It overlies folded rocks of the Anarchist Group and Kobau Formation with marked unconformity, but is itself only moderately tilted and warped. This is well shown by its distribution on the geologic map and indicates that the Ellemeham postdates major folding of the older rocks.

(2) The occurrence, in the upper member, of metamorphosed fragments of Anarchist and Kobau rocks indicates that it also postdates metamorphism and sub-sequent planation of the older rocks, although the Ellemeham itself is metamorphosed to greenschist facies, as noted above. Thus the Ellemeham contains evidence of a preceding and succeeding metamorphism. The younger metamorphism appears to be related to the intrusion of the Similkameen pluton, for shonkinite and malignite, lithologically identical to the malig-nitic border phase of the Similkameen, have intruded, hornfelsed, and fenitized (see section on Shankers Bend diatreme) the Ellemeham near the Similkameen River about a mile and a half south of Shankers Bend, and the weak metamorphism of the formation is attri-buted to this intrusive episode. Hence, the Ellemeham is older than the Similkameen, yet it is significantly younger than the metamorphism to which the older

rocks were subjected. Because the Ellemeham was deposited on rocks that probably were metamorphosed during the Late Triassic, is itself metamorphosed, and is overlain by unmetamorphosed rocks of early Eocene age, it is here assigned an age of Jurassic or Cretaceous.

PLUTONIC ROCKS

Except for serpentinite, the plutonic rocks, described below, are granitic rocks and are classified modally according to the scheme (modified after Johannsen, 1931) shown on figure 9. Modes were obtained from sawed slabs, generally 50 to 100 square centimeters in area, that were selectively stained (Laniz and others, 1964) to provide maximum distinction among the two feldspars, quartz, and mafic minerals. Modal analysis was made by superimposing on the slab a grid of black dots, almost microscopic in size, spaced to cover the entire area with a minimum of a thousand dots, and recording the mineral lying beneath each dot. The dot grids used were glass plate diapositives laid directly on the stained slab and viewed through a binocular steromicroscope. Results are plotted on figure 9; sample localities are shown on figures 10, 11, 12, and 13. The ratios of mafic and accessory minerals in the chemically analyzed specimens were estimated by point counting a single thin section viewed through a petrographic microscope. Results are listed on tables 3, 4, and 5.

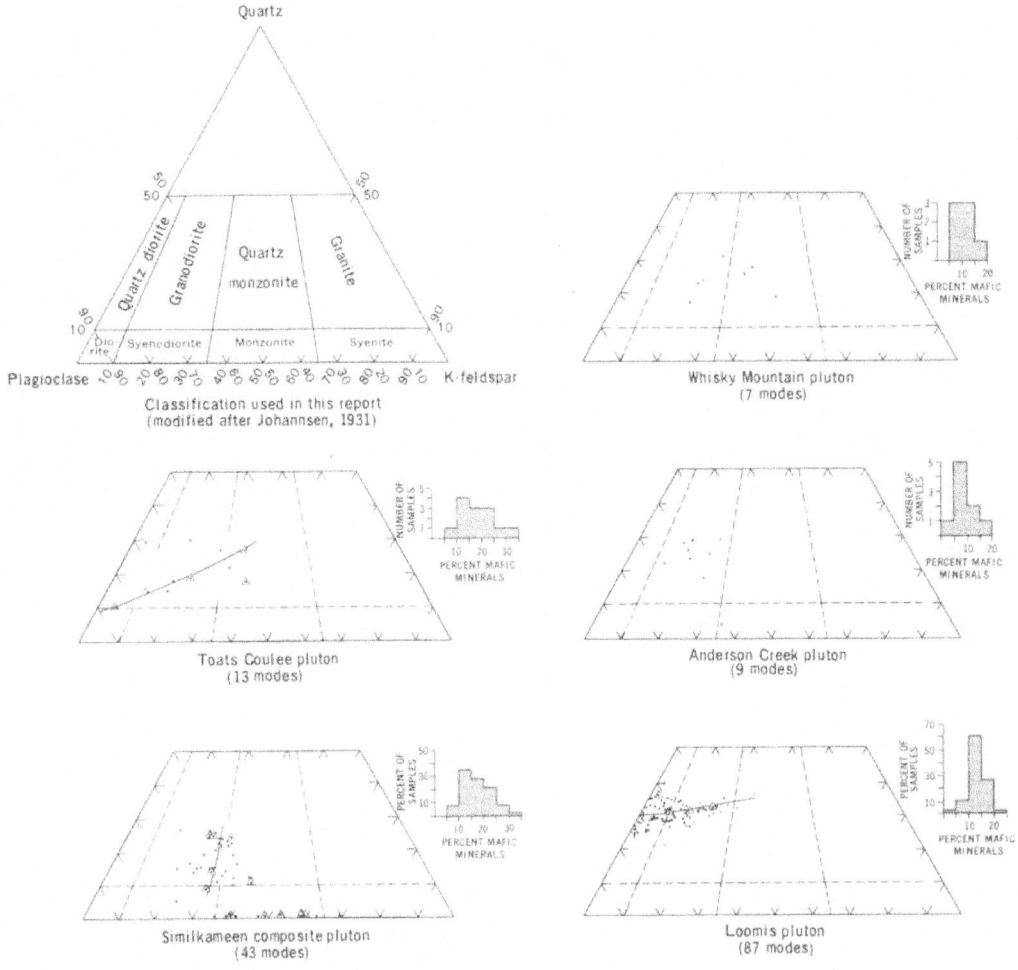

FIGURE 9.—Classification scheme and modal plots of quartz-rich granitic rocks from the Loomis quadrangle. Nomenclature of quartz-poor plutonic rocks discussed in text. Separate symbols indicate modes of chemically analyzed samples.

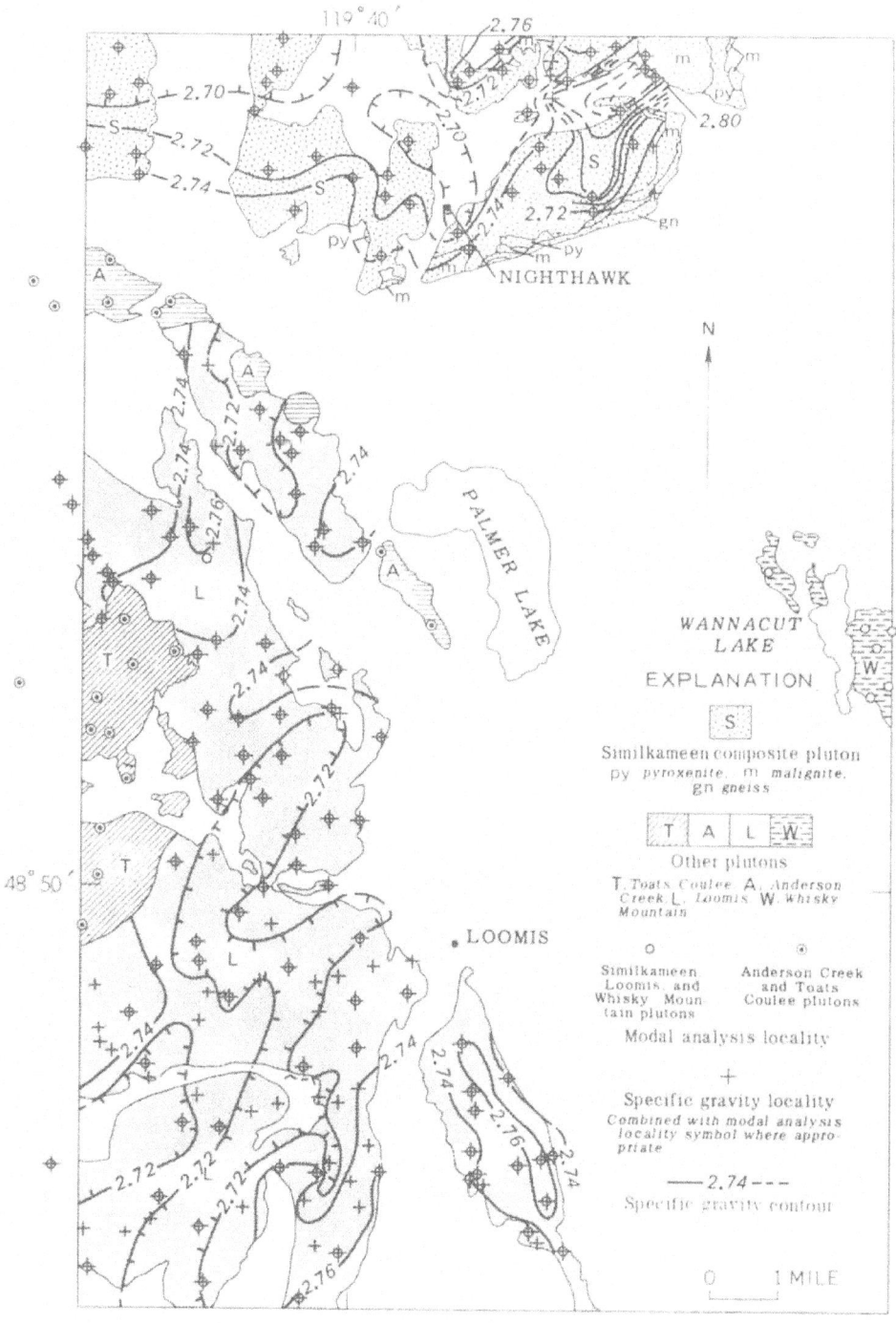

FIGURE 10.—Map of the Loomis quadrangle showing pluton outlines, sample localities, and contoured specific gravity data from the Loomis and Similkameen plutons.

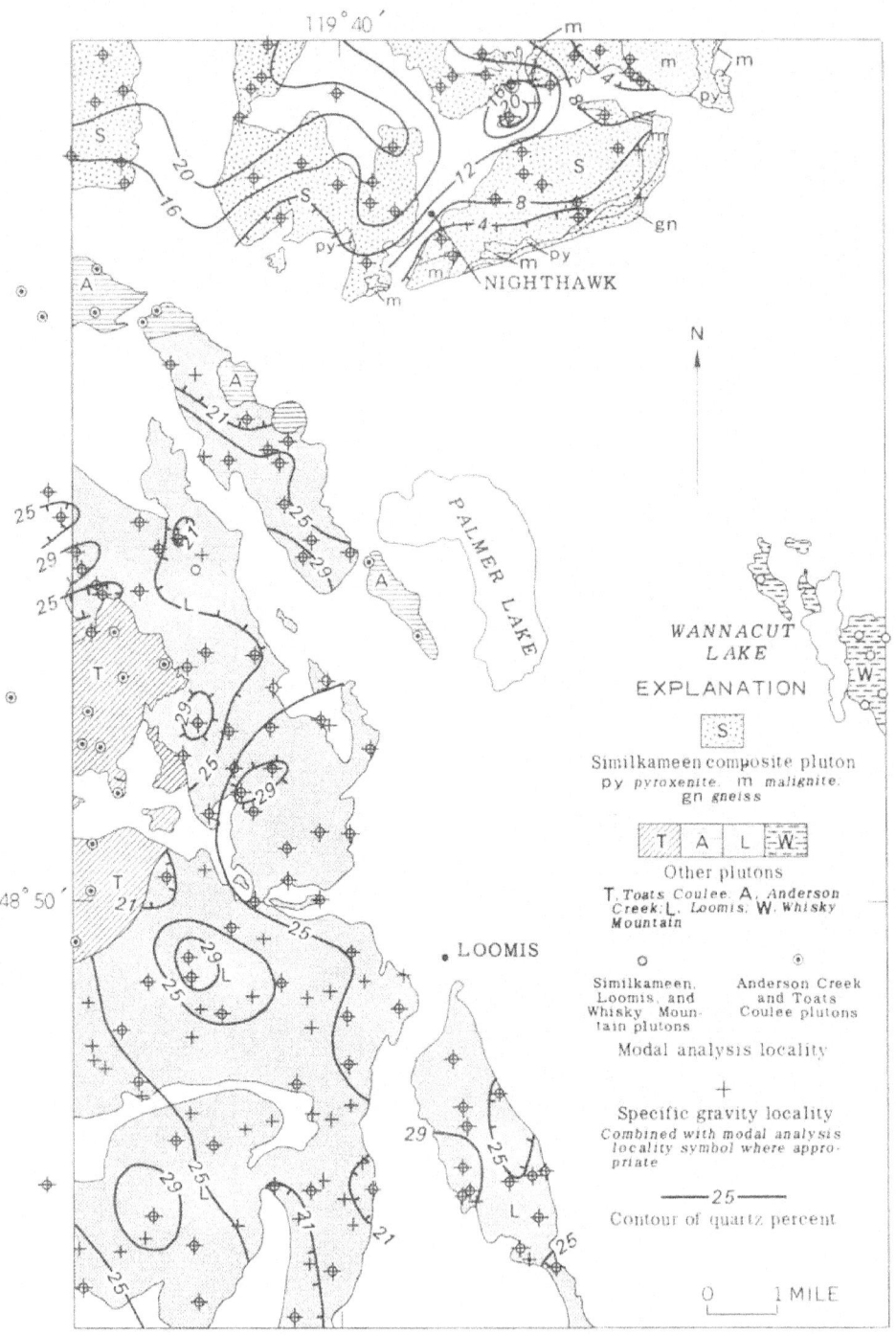

FIGURE 11.—Map of the Loomis quadrangle showing pluton outlines, sample localities, and contoured quartz content of the Loomis and Similkameen plutons.

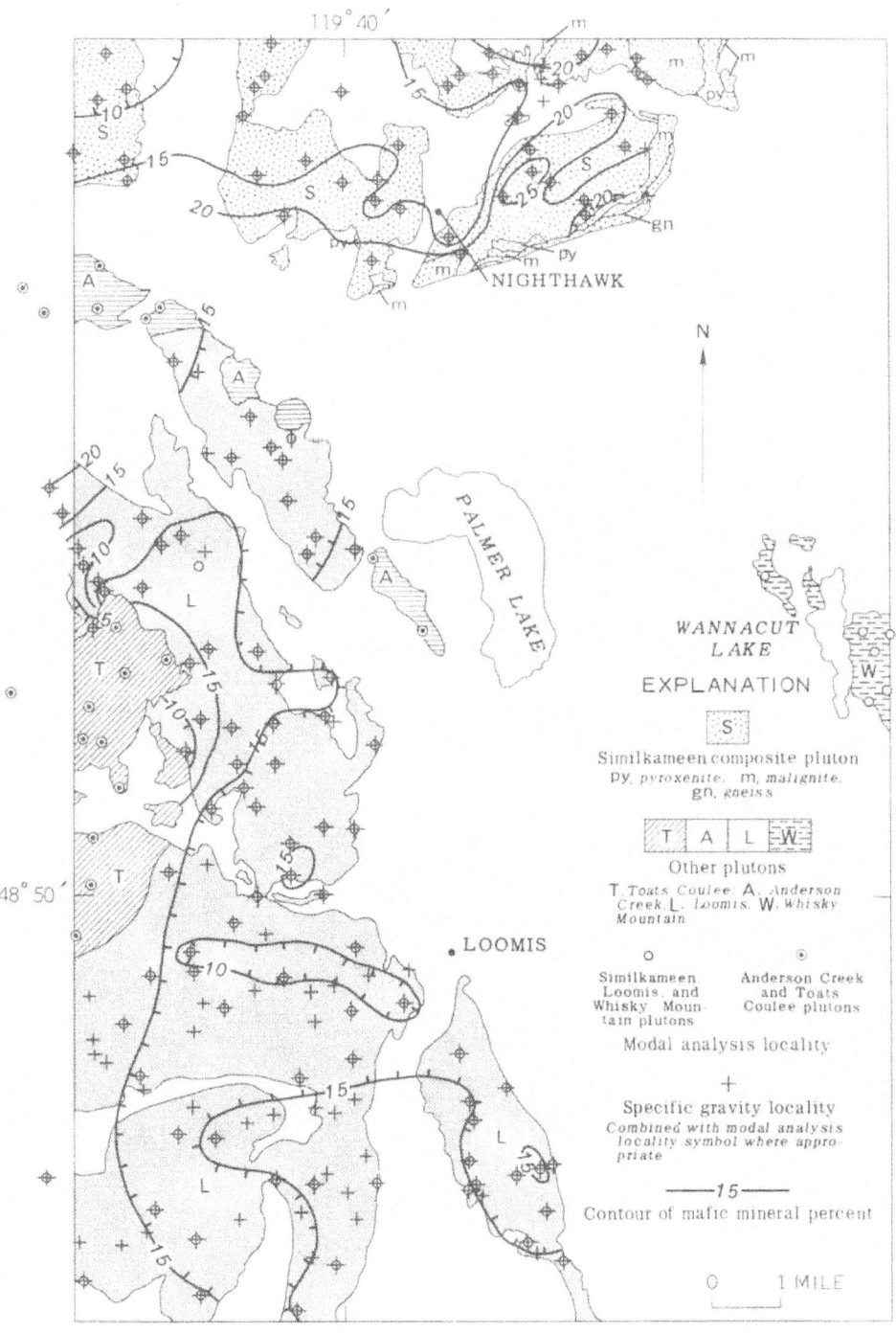

FIGURE 12.—Map of the Loomis quadrangle showing pluton outlines, sample localities, and contoured mafic mineral content of the Loomis and Similkameen plutons.

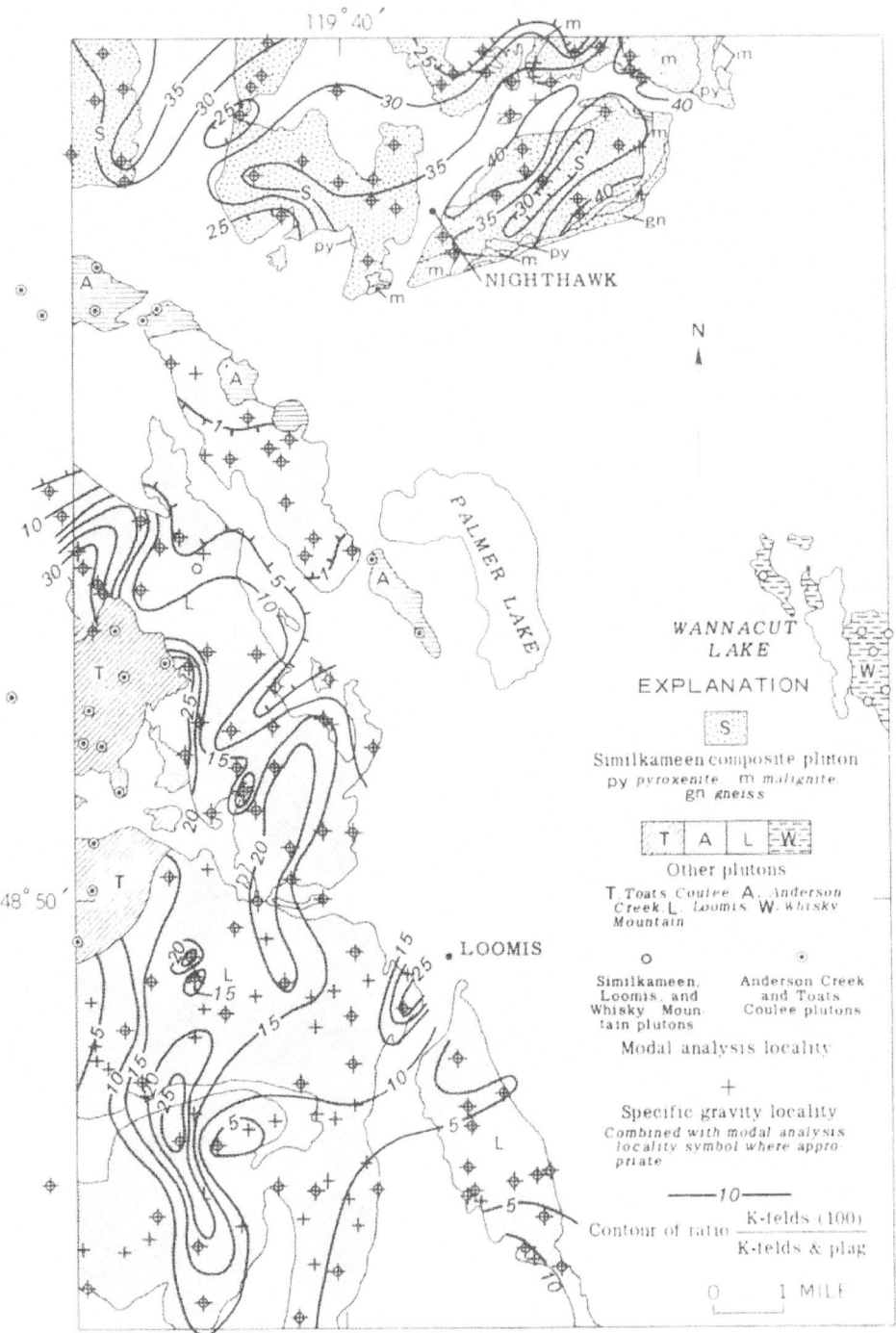

FIGURE 13.—Map of the Loomis quadrangle showing pluton outlines, sample localities, and K-feldspar as a percentage of total feldspar in the Loomis and Similkameen plutons.

TABLE 3.—Chemical and spectrographic analyses, norms, and modes of the Loomis and Toats Coulee plutons

	Loomis pluton					Toats Coulee pluton			
	L-502 [1] Quartz diorite	L-49-A Quartz diorite	L-342-A Quartz diorite	L-542 Granodiorite	L-425 Granodiorite	L-495-F-4 Quartz diorite	L-483-C Granodiorite	L-591 Quartz monzonite	L-433 Quartz monzonite
Chemical analyses [2]									
SiO_2	60.6	64.9	66.1	66.9	66.9	55.4	61.2	62.2	67.8
Al_2O_3	16.3	17.1	16.1	15.9	15.5	19.9	17.8	16.0	15.8
Fe_2O_3	2.7	2.1	1.5	2.0	1.8	3.6	2.4	1.9	1.5
FeO	2.9	2.1	1.9	1.9	1.8	3.8	2.6	3.8	1.7
MgO	2.2	1.6	1.5	1.3	1.3	2.3	1.6	2.0	0.93
CaO	5.8	4.7	4.5	4.0	4.0	6.2	3.9	4.1	2.8
Na_2O	3.6	3.7	4.0	3.4	3.5	4.5	4.4	3.4	3.5
K_2O	1.0	1.7	1.8	2.4	2.7	2.0	3.4	4.4	4.4
H_2O-	0.08	0.09	0.09	0.11	0.14	0.15	0.12	0.09	0.14
H_2O+	1.8	0.84	1.1	0.71	0.74	0.85	0.68	0.59	0.59
TiO_2	0.54	0.78	0.31	0.37	0.32	0.74	0.56	0.78	0.37
P_2O_5	0.18	0.14	0.10	0.11	0.12	0.56	0.24	0.25	0.14
MnO	0.15	0.14	0.11	0.12	0.12	0.14	0.09	0.13	0.08
CO_2	1.5	< .05	0.25	0.12	0.09	< .05	0.36	< .05	< .05
Total	99.35	99.89	99.36	99.34	99.03	100.14	99.35	99.64	99.75
Specific gravity (lump)	2.77	2.73	2.73	2.72	2.70	2.78	2.70	2.75	2.68
Specific gravity (powder)	2.78	2.77	2.78	2.72	2.73	2.81	2.73	2.74	2.66

Norms (weight percent)

Quartz	23.75	24.34	24.30	27.59	25.85	4.54	12.67	13.30	23.02
Corundum	2.67	0.96	0.21	0.98	0.04	0.00	1.21	0.00	0.53
Orthoclase	5.95	10.06	10.71	14.28	16.11	11.83	20.22	26.10	26.07
Albite	30.66	31.34	34.07	28.96	29.91	38.10	37.48	28.87	29.69
Anorthite	18.23	22.43	20.22	18.49	18.67	28.21	15.61	15.46	13.01
Wollastonite	0.00	0.00	0.00	0.00	0.00	0.09	0.00	1.39	0.00
Enstatite	5.52	3.99	3.76	3.26	3.27	5.73	4.01	5.00	2.32
Ferrosilite	2.50	1.10	1.96	1.46	1.53	3.04	2.05	4.38	1.42
Magnetite	3.94	3.05	2.19	2.92	2.64	5.22	3.50	2.77	2.18
Ilmenite	1.03	1.48	0.59	0.71	0.61	1.41	1.07	1.49	0.70
Apatite	0.43	0.33	0.24	0.26	0.29	0.85	0.57	0.59	0.33
Calcite	3.43	0.00	0.57	0.28	0.21	0.00	0.82	0.00	0.00
Total	98.11	99.08	98.82	99.19	99.13	99.02	99.21	99.35	99.27

Modes

Quartz	21	27	31	27	28	8	15	13	24
Microperthitic microcline	...	7	2	10	16	3	14	28	27
Plagioclase	60	54	54	49	44	68	58	37	39
An, range shown where zoned	(25–50)	(25–45)	(25–45)	(20–50)	(30–40)	(25–40)	(25–40)	(30–35)	(25–30)
Biotite	...	4	9	10	7	10	10	10	8
Hornblende	...	6	<1	1	1	9	...	9	1
Epidote	3	1	3	<1	1	<1	...	<1	<1
Chlorite	15	<1	1	1	1	<1	1	1	<1
Magnetite and ilmenite	1	1	<1	1	2	2	2	1	1
Sphene	<1	<1	<1	<1	<1	<1	<1	1	1
Apatite	<1	<1	<1	1	<1	<1	<1	1	<1

(See footnotes at end of table.)

TABLE 3.—Chemical and spectrographic analyses, norms, and modes of the Loomis and Toats Coulee plutons—Continued

Semiquantitative spectrographic analyses for minor elements [3]
(C. Heropoulos, analyst)

| | Loomis pluton | | | | | Toats Coulee pluton | | | |
	L-502 [1] Quartz diorite	L-49-A Quartz diorite	L-342-A Quartz diorite	L-542 Granodiorite	L-425 Granodiorite	L-495-F-4 Quartz diorite	L-483-C Granodiorite	L-591 Quartz monzonite	L-433 Quartz monzonite
B	0.001	0	0.001	0	0.001	0	0	0.005	0.001
Ba	0.15	0.2	0.2	0.15	0.2	0.15	0.2	0.15	0.2
Be	0	0	0	0	0	0.00015	0.0002	0.0002	0.0002
Ce	0	0	0	0	0	0	0	0	0
Co	0.0015	0.001	0.001	0.001	0.0007	0.002	0.001	0.002	0.0007
Cr	0.003	0.0015	0.0015	0.001	0.0007	0.001	0.0007	0.002	0.0005
Cu	0.0007	0.001	0.0005	0.0007	0.0007	0.003	0.003	0.01	0.001
Ga	0.0015	0.0015	0.0015	0.0015	0.0015	0.002	0.002	0.002	0.0015
La	0	0	0	0	0	0.003	0.005	0.007	0.005
Nb	0	0	0	0	0	0.001	0.001	0.002	0.001
Ni	0.001	0.0005	0.0005	0.0005	0.0003	0.0007	0.0003	0.0015	0.0002
Pb	0	0	0	0	0	0	0	0	0.002
Sc	0.002	0.0015	0.0015	0.001	0.001	0.002	0.0015	0.002	0.001
Sr	0.07	0.05	0.05	0.03	0.03	0.15	0.1	0.07	0.07
V	0.015	0.01	0.01	0.01	0.007	0.015	0.01	0.015	0.007
Y	0.002	0.002	0.002	0.0015	0.0015	0.003	0.003	0.005	0.002
Yb	0.0002	0.0002	0.0002	0.00015	0.0002	0.0003	0.0003	0.0005	0.0002
Zr	0.015	0.015	0.01	0.01	0.01	0.03	0.03	0.05	0.02
Nd	....	....	....	....	....	0	0	0	0

1/ Sample locations: L-502, SE$\frac{1}{4}$ sec. 5, T. 39 N., R. 25 E.

L-498-A, SE$\frac{1}{4}$ sec. 34, T. 39 N., R. 25 E.

L-342-A, NE$\frac{1}{4}$ sec. 24, T. 38 N., R. 25 E.

L-542, NW$\frac{1}{4}$ sec. 9, T. 38 N., R. 25 E.

L-425, SE$\frac{1}{4}$ sec. 17, T. 39 N., R. 25 E.

L-495-F-4, SE$\frac{1}{4}$ sec. 17, T. 39 N., R. 25 E.

L-483-C, NE$\frac{1}{4}$ sec. 18, T. 39 N., R. 25 E.

L-591, SW$\frac{1}{4}$ sec. 13, T. 39 N., R. 24 E.

L-433, NW$\frac{1}{4}$ sec. 19, T. 39 N., R. 25 E.

2/ Chemical analyses by P. Elmore, S. Botts, G. Chloe, J. Glenn, L. Artis, H. Smith, and D. Taylor, using the rapid method of Shapiro and Brannock (1956).

3/ Results are reported in percent to the nearest number in the series 1, 0.7, 0.5, 0.3, 0.2, 0.15, and 0.1, etc., which represent approximate midpoints of interval data on a geometric scale. The assigned interval for semi-quantitative results will include the quantitative value about 30 percent of the time.

TABLE 4.—Chemical and spectrographic analyses,

	L-507-B[1]/ Pyroxenite	L-275-C Pyroxenite	L-569 Shonkinite	L-301 Shonkinite	L-504-B Shonkinite	L-442 Syenodiorite

Chemical

SiO_2	39.7	40.3	49.0	52.1	53.1	56.3
Al_2O_3	7.0	11.3	14.5	14.9	16.8	15.9
Fe_2O_3	10.2	7.3	4.7	2.5	4.7	4.3
FeO...............	8.9	10.0	6.1	5.6	4.1	4.4
MgO	9.5	7.7	4.7	5.5	3.0	3.1
CaO	18.7	14.5	10.2	7.7	7.2	6.9
Na_2O	0.76	2.0	3.0	3.2	3.1	3.4
K_2O...............	0.80	1.7	4.1	5.7	5.3	3.0
H_2O-..............	0.08	0.13	0.15	0.13	0.15	0.16
H_2O+..............	0.72	0.97	0.67	0.87	0.73	0.71
TiO_2	1.6	1.6	1.0	0.65	0.80	0.90
P_2O_5	1.6	1.5	0.81	0.67	0.63	0.52
MnO	0.34	0.37	0.24	0.18	0.20	0.22
CO_2	< .05	0.15	< .05	0.08	< .05	< .05
Total	99.90	99.52	99.17	99.78	99.81	99.81
Specific gravity ...... (lump)	3.42	3.30	2.88	2.91	2.87	2.85
Specific gravity ...... (powder)	. . .	. . .	3.05	2.90	. . .	. . .

Norms

Quartz	0.00	0.00	0.00	0.00	0.00	8.05
Corundum	0.00	0.00	0.00	0.00	0.00	0.00
Orthoclase	0.00	10.09	24.43	33.76	31.38	17.76
Albite	0.00	0.38	14.10	14.14	26.15	28.83
Anorthite	13.34	16.92	14.11	9.48	16.30	19.30
Leucite	3.71	0.00	0.00	0.00	0.00	0.00
Nepheline..........	3.49	9.00	6.23	7.04	0.07	0.00
Wollastonite	28.17	18.61	13.19	9.98	6.41	4.84
Enstatite	20.46	11.39	8.16	6.10	4.35	7.74
Ferrosilite	5.11	6.17	4.26	3.33	1.57	3.46

(See footnotes at end of table.)

norms, and modes of the Similkameen composite pluton

L209-A Shonkinite gneiss	L-233 Monzonite	L-235-A Quartz monzonite	L-283 Granodiorite	L-618 Granodiorite	L-589-C Quartz monzonite	L-281 Granodiorite	L-376-B Granodiorite

analyses [2]

L209-A Shonkinite gneiss	L-233 Monzonite	L-235-A Quartz monzonite	L-283 Granodiorite	L-618 Granodiorite	L-589-C Quartz monzonite	L-281 Granodiorite	L-376-B Granodiorite
57.1	60.0	60.4	62.2	65.2	65.5	66.2	67.8
20.4	19.1	15.9	17.1	16.7	16.1	16.1	16.2
3.5	2.0	2.7	2.5	2.4	2.1	1.9	1.5
1.0	2.0	3.3	2.6	1.7	1.8	1.7	1.6
0.9	1.5	2.5	1.9	1.4	1.4	1.1	0.9
3.5	3.9	4.9	5.0	4.1	4.0	3.9	3.4
4.2	5.7	3.7	4.0	3.9	3.9	3.9	4.0
7.5	4.5	4.5	3.4	3.1	3.5	3.6	3.5
0.10	0.08	0.19	0.10	0.05	0.04	0.11	0.11
0.49	0.34	0.60	0.32	0.52	0.64	0.60	0.41
0.46	0.47	0.63	0.51	0.42	0.38	0.37	0.32
0.21	0.19	0.38	0.29	0.18	0.19	0.18	0.15
0.14	0.12	0.14	0.13	0.13	0.12	0.13	0.10
0.21	< .05	< .05	< .05	< .05	0.85	0.06	< .05
99.71	99.90	99.84	100.05	99.80	100.52	99.85	99.99
. . .	2.69	2.75	2.74	2.71	2.71	2.71	2.69
. . .	. . .	. . .	. . .	2.72	2.72	. . .	. . .

(weight percent)

L209-A Shonkinite gneiss	L-233 Monzonite	L-235-A Quartz monzonite	L-283 Granodiorite	L-618 Granodiorite	L-589-C Quartz monzonite	L-281 Granodiorite	L-376-B Granodiorite
0.00	0.00	8.93	13.15	20.07	19.43	20.34	22.32
0.00	0.00	0.00	0.00	0.00	0.00	0.00	0.01
44.45	26.62	26.63	20.08	18.36	20.74	21.31	20.69
29.00	48.28	31.36	33.83	33.07	33.09	33.05	33.85
14.70	13.25	13.51	18.65	18.94	16.13	15.82	15.89
0.00	0.00	0.00	0.00	0.00	0.00	0.00	0.00
3.60	0.00	0.00	0.00	0.00	0.00	0.00	0.00
0.00	2.03	3.49	1.77	0.11	0.92	0.84	0.00
0.00	3.61	6.24	4.73	3.49	3.50	2.74	2.24
0.00	1.42	3.05	2.11	0.69	1.17	1.19	1.36

TABLE 4.—Chemical and spectrographic analyses, norms,

	L-507-B[1/] Pyroxenite	L-275-C Pyroxenite	L-569 Shonkinite	L-301 Shonkinite	L-504-B Shonkinite	L-442 Syenodiorite

Norms (weight

Forsterite	2.26	5.52	2.56	5.35	2.20	0.00
Fayalite	0.62	3.29	1.47	3.22	0.88	0.00
Calcium orthosilicate	0.50	0.00	0.00	0.00	0.00	0.00
Magnetite	14.81	10.64	6.87	3.63	6.83	6.25
Hematite............	0.00	0.00	0.00	0.00	0.00	0.00
Ilmenite	3.04	3.05	1.92	1.24	1.52	1.71
Apatite.............	3.79	3.57	1.94	1.59	1.50	1.23
Calcite.............	0.00	0.34	0.00	0.18	0.00	0.00
Total	99.30	98.97	99.24	99.04	99.16	99.17

Modes

Quartz	...	...	...	...	...	7
Microline............ (micro- perthitic)	...	2	27	25	26	22
Plagioclase An, range shown where zoned.........	... ...	< 1 olig(?)	18 (30–45)	23 (20–25)	27 (20–30)	48 (20–35)
Biotite	10	< 1	1	15	< 1	1
Muscovite............	...	...	1	...	< 1	...
Hastingsite or hornblende..........	19	53	30	16	24	13
Clinopyroxene	61	39	18	16	9	< 1
Epidote..............	< 1	< 1	1	...	12	8
Sphene	< 1	< 1	< 1	< 1	< 1	1
Magnetite and ilmenite	6	1	2	< 1	1	1
Apatite..............	3	3	2	1	1	< 1
Nepheline	...	...	< 1	3	...	...
Zeolites.............	...	...	< 1	...	...	...

Semiquantitative spectrographic (C. Heropoulos,

B	0	0	0.0015	0.005	0	0.0015
Ba	0.07	0.07	0.2	0.15	0.2	0.15
Be	0.0002	0.0003	0.0002	0.0005	0.0002	0.003
Ce	0.02	0.02	0	0	0.015	0

(See footnotes at end of table.)

and modes of the Similkameen composite pluton—Continued

L209-A Shonkinite gneiss	L-233 Monzonite	L-235-A Quartz monzonite	L-283 Granodiorite	L-618 Granodiorite	L-589-C Quartz monzonite	L-281 Granodiorite	L-376-B Granodiorite

percent) — Continued

1.57	0.09	0.00	0.00	0.00	0.00	0.00	0.00
0.00	0.04	0.00	0.00	0.00	0.00	0.00	0.00
0.00	0.00	0.00	0.00	0.00	0.00	0.00	0.00
2.35	2.90	3.92	3.62	3.49	3.05	2.74	2.18
1.89	0.00	0.00	0.00	0.00	0.00	0.00	0.00
0.88	0.89	1.20	0.97	0.80	0.72	0.70	0.61
0.50	0.45	0.90	0.69	0.43	0.45	0.43	0.36
0.48	0.00	0.00	0.00	0.00	0.11	0.14	0.00
99.42	99.58	99.23	99.60	99.45	99.31	99.32	99.51

...	...	8	12	18	20	22	23
32	35	28	22	22	24	20	19
39	54	36	48	45	41	46	49
ab-olig	(20-25)	20	(20-35)	(30-35)	(25-30)	(20-35)	(20-35)
9	< 1	1	3	5	6	3	3
11	...	< 1	...	...	...	...	...
...	7	19	10	8	4	3	4
...	1	...	...	...	...	...	...
6	1	4	3	3	4	4	1
1	1	< 1	1	< 1	< 1	1	< 1
1	1	1	1	1	1	< 1	1
< 1	< 1	2	< 1	< 1	1	< 1	< 1
...	...	...	...	...	...	...	...
...	...	...	...	...	...	...	...

analyses for minor elements [3/]
analyst)

0.0015	0	0.007	0.001	0	0	0.002	0
0.15	0.2	0.15	0.2	0.15	0.15	0.15	0.15
0.0007	0.0003	0.0005	0.0002	0.00015	0.0002	0.0002	0.0002
0	0	0	0	0	0	0	0

TABLE 4.—Chemical and spectrographic analyses, norms,

	L-507-B[1] Pyroxenite	L-275-C Pyroxenite	L-569 Shonkinite	L-301 Shonkinite	L-504-B Shonkinite	L-442 Syenodiorite

Semiquantitative spectrographic analyses
(C. Heropoulos,

Co	0.01	0.007	0.005	0.003	0.005	0.003
Cr	0.01	0.003	0.005	0.02	0.001	0.002
Cu	0.015	0.05	0.03	0.02	0.005	0.005
Ga	0.002	0.002	0.0015	0.0015	0.002	0.002
La	0.01	0.015	0.007	0.007	0.007	0.007
Nb	0.0015	0.003	0.001	0.001	0	0.002
Ni	0.007	0.003	0.002	0.01	0.0015	0.0015
Pb	0	0	0	0.0015	0.001	0.0015
Sc	0.015	0.01	0.005	0.003	0.003	0.003
Sr	0.07	0.1	0.2	0.2	0.2	0.15
V	0.15	0.1	0.05	0.03	0.07	0.05
Y	0.007	0.007	0.003	0.002	0.003	0.005
Yb	0.0007	0.0007	0.0003	0.0003	0.0003	0.0005
Zr	0.02	0.02	0.015	0.015	0.01	0.02
Nd	0.015	0.015	0	0	0.01	0

[1] Sample locations: L-507-B, SW$\frac{1}{4}$ sec. 18, T. 40 N., R. 26 E.

L-569 and L-275-C, NE$\frac{1}{4}$ sec. 4, T. 40 N., R. 26 E.

L-301, SE$\frac{1}{4}$ sec. 3, T. 40 N., R. 26 E.

L-504-B, NE$\frac{1}{4}$ sec. 9, T. 40 N., R. 26 E.

L-442, NE$\frac{1}{4}$ sec. 8, T. 40 N., R. 26 E.

L-209-A, SW$\frac{1}{4}$ sec. 17, T. 40 N., R. 26 E.

L-233, SE$\frac{1}{4}$ sec. 13, T. 40 N., R. 25 E.

L-235-A, SE$\frac{1}{4}$ sec. 7, T. 40 N., R. 26 E.

L-283, NE$\frac{1}{4}$ sec. 15, T. 40 N., R. 25 E.

L-618, SE$\frac{1}{4}$ sec. 11, T. 40 N., R. 25 E.

L-589-C, NE$\frac{1}{4}$ sec. 7, T. 40 N., R. 26 E.

L-281, SE$\frac{1}{4}$ sec. 3, T. 40 N., R. 25 E.

L-376-B, SE$\frac{1}{4}$ sec. 6, T. 40 N., R. 25 E.

and modes of the Similkameen composite pluton—Continued

L209-A Shonkinite gneiss	L-233 Monzonite	L-235-A Quartz monzonite	L-283 Granodiorite	L-618 Granodiorite	L-589-C Quartz monzonite	L-281 Granodiorite	L-376-B Granodiorite

for minor elements[3]—Continued
analyst)

L209-A Shonkinite gneiss	L-233 Monzonite	L-235-A Quartz monzonite	L-283 Granodiorite	L-618 Granodiorite	L-589-C Quartz monzonite	L-281 Granodiorite	L-376-B Granodiorite
0.001	0.0007	0.003	0.0015	0.0007	0.001	0.001	0.0007
0.0007	0.0015	0.005	0.0015	0.0007	0.001	0.0007	0.0007
0.002	0.001	0.03	0.002	0.001	0.0005	0.001	0.00015
0.003	0.002	0.002	0.0015	0.0015	0.0015	0.002	0.0015
0.005	0.003	0.005	0.005	0.003	0.005	0	0
0.0015	0.0015	0.0015	0.0015	0.001	0.001	0	0.001
0.0005	0.0007	0.003	0.0007	0.0002	0.0003	0.0005	0.0005
0.007	0.002	0.002	0.002	0.001	0.002	0.003	0.002
0.001	0.001	0.002	0.002	0.001	0.001	0.001	0.0007
0.2	0.15	0.15	0.15	0.07	0.1	0.1	0.1
0.015	0.015	0.03	0.02	0.01	0.01	0.015	0.01
0.003	0.003	0.003	0.003	0.002	0.002	0.002	0.0015
0.0003	0.0003	0.0003	0.0003	0.0002	0.0002	0.0002	0.0002
0.015	0.02	0.03	0.015	0.015	0.01	0.01	0.01
0	0	0	0	0	0	...	...

[2] Analysts for L-618 and L-589-C: P. Elmore, L. Artis, G. Chloe, J. Glenn, H. Smith, D. Taylor; for L-301 and L-569: P. Elmore, L. Artis, S. Botts, G. Chloe, J. Glenn, H. Smith, D. Taylor; for all others: P. Elmore, S. Botts, L. Artis (using rapid method of Shapiro and Brannock, 1956).

[3] Results are reported in percent to the nearest number in the series 1, 0.7, 0.5, 0.3, 0.2, 0.15, and 0.1, etc., which represent approximate midpoints of interval data on a geometric scale. The assigned interval for semiquantitative results will include the quantitative value about 30 percent of the time.

TABLE 5.—Chemical and spectrographic analyses, norms, and modes of rocks from the Shankers Bend diatreme and greenstone from the Ellemeham Formation

	L-620-P[1] Shonkinite	L-444-K Fenitized greenstone	L-526-A Greenstone
Chemical analyses[2]			
SiO_2	49.9	44.9	48.2
Al_2O_3	16.0	16.8	13.3
Fe_2O_3	4.8	2.7	1.5
FeO.............	4.5	10.0	7.8
MgO............	4.2	4.4	6.2
CaO	9.1	8.4	7.2
Na_2O	3.9	4.0	3.5
K_2O	3.9	2.3	0.22
H_2O-	0.11	0.25	0.24
H_2O+	1.2	2.4	4.2
TiO_2	0.76	2.9	1.8
P_2O_5	0.62	0.63	0.25
MnO	0.25	0.23	0.13
CO_2	0.05	0.08	5.3
Total	99.29	99.99	99.84
Specific gravity .... (lump)	2.93	2.98	2.78
Specific gravity .... (powder)	2.93	3.00	2.82
Norms (weight percent)			
Quartz	0	0	12.81
Corundum	0	0	7.10
Orthoclase	23.21	13.59	1.30
Albite	19.11	17.68	29.66
Anorthite	14.74	21.10	0.58

(See footnotes at end of table.)

TABLE 5.—Chemical and spectrographic analyses, norms, and modes of rocks from the Shankers Bend diatreme and greenstone from the Ellemeham Formation—Continued

	L-620-P[1] Shonkinite	L-444-K Fenitized greenstone	L-526-A Greenstone
Norms (weight percent)—Continued			
Nepheline	7.65	8.76	0
Wollastonite.......	11.00	6.66	0
Enstatite	7.57	3.17	15.47
Ferrosilite	2.54	3.40	10.37
Forsterite	2.08	5.46	0
Fayalite	0.77	6.46	0
Magnetite	7.01	3.92	2.18
Ilmenite	1.45	5.51	3.42
Apatite	1.48	1.49	0.59
Calcite...........	0.12	0.18	12.07
Total	98.73	97.38	95.55
Modes[3] (X = major constituent)			
Quartz	. . .	. . .	X
K-feldspar	23	5	. . .
Plagioclase........ An	22 30	20 albite	X albite?
Biotite	2	40	. . .
Hastingsite	35	6	. . .
Clinopyroxene	9	6	. . .
Sphene	<1	<1	. . .
Apatite...........	1	<1	. . .
Opaque minerals ...	<1	<1	. . .
Nepheline (including inferred pseudo-morphous zeolites)	5	2	. . .

(See footnotes at end of table.)

TABLE 5.—Chemical and spectrographic analyses, norms, and modes of rocks from the Shankers Bend diatreme

and greenstone from the Ellemeham Formation—Continued

	L-620-P[1] Shonkinite	L-444-K Fenitized greenstone	L-526-A Greenstone
Modes [3]—Continued (X-major constituent)			
Garnet	2	. . .	. . .
Chlorite	. . .	. . .	X
Semiquantitative spectrographic analyses for minor elements [4]			
Ba	0.15	0.1	0.015
Be	0.0007	0.0003	0
Co	0.003	0.005	0.005
Cr	0.002	0.001	0.07
Cu	0.007	0.015	0.01
Ga	0.0015	0.002	0.0015
La	0.007	0.005	0
Nb	0.001	0.005	0.0015
Ni	0.0015	0.005	0.02
Sc	0.005	0.003	0.005
Sr	0.2	0.1	0.03
V..............	0.03	0.03	0.02
Y..............	0.003	0.005	0.003
Yb	0.0003	0.0005	0.0003
Zr..............	0.02	0.03	0.015

[1] Sample locations: L-620-P and L-444-K, SE$\frac{1}{4}$ sec. 14, T. 40 N., R. 26 E.
L-526-A, NE$\frac{1}{4}$ sec. 1, T. 39 N., R. 26 E.

[2] Chemical analyses by P. Elmore, L. Artis, S. Botts, G. Chloe, J. Glenn, H. Smith, D. Taylor, using the rapid method of Shapiro and Brannock (1956).

[3] Modes of L-444-K and L-526-A estimated from examination of thin section and X-ray powder diffractogram.

[4] C. Heropoulos, analyst. Results are reported in percent to the nearest number in the series 1, 0.7, 0.5, 0.3, 0.2, 0.15, and 0.1, etc., which represent approximate midpoints of interval data on a geometric scale. The assigned interval for semiquantitative results will include the quantitative value about 30 percent of the time.

SERPENTINITE

Narrow elongate ribs and irregular masses of serpentinized ultramafic rock crop out at several localities in the western half of the quadrangle. On fresh exposures the ultramafics are typically a massive, compact, fine- to medium-grained dark- to dark-bluish-gray rock. Exterior surfaces weather to a porous brownish-orange rind about a quarter of an inch thick, with an irregularly corrugated "elephant-hide" texture. Thin (1/16 to 1/4 inch thick) curving seams of light-green chrysotile asbestos are widespread but not abundant. Very thin irregular dark-gray chromite-bearing laminations were also observed. The laminations customarily parallel the long dimension of the elongate masses. The rock is magnetic, resulting in compass deviations at the outcrops.

The elongate body on the southeast slope of Chopaka Mountain (secs. 29 and 32, T. 40 N., R. 25 E.) and the irregular mass on Grandview Mountain (sec. 14, T. 39 N., R. 25 E.) are serpentinites, composed of antigorite with subordinate carbonate and magnetite, locally with accessory chromite and pyrite. The mass on Little Chopaka Mountain (sec. 16, T. 40 N., R. 25 E.) is also a serpentinite, but contains both relict olivine and relict enstatite embedded in a matrix of fibrous antigorite; accessory minerals are magnetite, chromite, and talc.

Tremolite and diopside are abundant within the serpentinite in the vicinity of the contact with the Loomis pluton on Chopaka Mountain. The tremolite occurs as decussate, acicular or fibrous grains, and the diopside as irregular anhedral poikiloblastic grains. Both are superimposed on a decussate mat of needlelike or sheaflike antigorite and presumably are the result of metamorphism of the serpentinite during the intrusion of the pluton.

Slickensides and gouge mark the eastern contact of the serpentinite on the southeast slope of Chopaka Mountain with the rocks of the Kobau Formation. The western contact is covered, but probably is sharp. The serpentinite contains a few large elongate inclusions of country rock, such as quartzite similar to that of the adjacent Kobau. The mass must have been emplaced prior to the intrusion of the Loomis pluton, judging by

the previously noted metamorphism of the serpentinite by the pluton, but subsequent to the folding of the metamorphic rocks (see Structural Geology), and thus is approximately mid-Triassic in age.

The serpentinite on the southeast slope of Chopaka Mountain is part of a family which includes other masses on Chopaka Mountain west of the quadrangle boundary. The group was first mapped by Smith and Calkins and designated serpentinite and pyroxenite by them (1904, p. 47, p. 73). Subsequently Daly (1912, p. 430) mapped the northern part of the belt and evidently found considerable relict olivine in the ultramafic rocks, as he concluded that they were serpentinized dunite (Daly, 1912, p. 434). Considering the opinion of these previous workers in conjunction with the results of our own work, it is probable that these ultramafics, as well as the others of the Loomis quadrangle, represent serpentinized rocks of the peridotite-dunite clan.

Petrologically, the serpentinite in the Loomis quadrangle belong to a group of similar rocks scattered along a wide belt through the Pacific Coast States and British Columbia (Hess, 1939). The ultramafic rocks of this belt include both the "alpine-type" of Hess and the "zoned ultramafic complexes" of Noble and Taylor (1960). The elongate outline of the majority of the Loomis serpentinites, together with their small size and lack of zoning, suggests to us that they are of the "alpine-type."

WHISKY MOUNTAIN PLUTON

The Whisky Mountain pluton, composed of porphyritic granodiorite and quartz monzonite, lies athwart the eastern boundary of the Loomis quadrangle, occupying an area of about 6 square miles, 80 percent of which is in the adjacent Oroville quadrangle. It was termed the "Whisky Mountain stock" by Waters and Krauskopf (1941, p. 1370-1371) for the low, rounded mountain it underlies east of Wannacut Lake in the Oroville quadrangle.

Compositions of seven modally analyzed specimens from the square mile or so in the Loomis quadrangle are shown on figure 9. Plagioclase (An$_{10-20}$)

is only weakly zoned, but zonal distribution of saussurite indicates more pronounced original zoning. Albitization is indicated by the presence of rims and veinlets of albite. Mafic minerals—chiefly biotite, chlorite, epidote, and magnetite—are commonly intergrown in shapes suggesting former hornblende crystals; a few hornblende relicts are locally preserved. Sphene is a megascopically visible accessory. Most of the rock is porphyritic, showing conspicuous euhedral, somewhat poikilitic phenocrysts of microcline microperthite, the largest of which measure 2 inches in maximum dimension. Sawed and selectively stained specimens reveal that plagioclase forms abundant subhedral to euhedral phenocrysts distinctly smaller than the microcline, but significantly larger than the grain size of the matrix. Microcline in the matrix is anhedral and interstitial to the plagioclase. In several specimens quartz is clustered in aggregates comparable in size to the plagioclase phenocrysts. Textures of the nonporphyritic rocks, and of the matrix in porphyritic varieties, are generally hypautomorphic-granular. As pointed out by Waters and Krauskopf (1941), the rock is structureless except for local foliated zones near contacts.

Beyond the fact that the Whisky Mountain pluton discordantly cuts rocks of the Spectacle Formation, its age is unknown. In exposures east of the quadrangle near the Okanogan River, rocks of the pluton show marked cataclastic deformation possibly caused by the intrusion of the Colville batholith. This suggests that the Whisky Mountain pluton is at least not the youngest of the Mesozoic plutons and it is therefore tentatively assigned an age of Triassic or Jurassic.

LOOMIS PLUTON

The Loomis pluton is composed of rocks whose compositions span the quartz diorite and granodiorite fields shown in figure 9, and whose average bulk composition lies near the boundary between these fields. The name "Loomis" was first applied to this unit by Pelton (1957) for the part of the pluton that is exposed east of the Sinlahekin Valley in Aeneas Mountain. Hibbard (1962) later applied the name "Gold Hill" to

the part of the pluton that is west of the Sinlahekin-Palmer Lake drainage, the eastern boundary of his mapping, and he was thus uncertain as to whether or not the two are correlative. There is now little doubt that they are correlative, and we have adopted Pelton's name for the entire pluton because the name is appropriate and predates Hibbard's work. Hibbard's (1962) map shows that the pluton occupies about 15 square miles west of the Loomis quadrangle boundary.[1]

The modes in table 3 are considered to be representative of that part of the pluton within the quadrangle. The rock is typically medium grained and the dominant texture is hypautomorphic-granular (fig. 14). Conspicuous books of euhedral biotite and discrete, somewhat scattered, broad subhedral hornblende prisms are characteristic megascopic textural features. Quartz shows a tendency to occur in aggregates rather than as disseminated grains, thus lending an additional distinctive element to the texture. Anhedral, poikilitic phenocrysts of K-feldspar are sparsely distributed through the southern half of the pluton. The phenocrysts are typically 5 to 15 mm. long, and locally number as many as 15 per 10,000 sq. mm. of surface area. The phenocrysts are unique in that the volume ratio of host crystal or oikocryst to the poikilitically included grains commonly is less than 1:1. As a result the oikocrysts are quite inconspicuous, and except where their presence is revealed by occasional reflections from cleavage surfaces, the texture of the rock resembles hypautomorphic granular. Chlorite, altered from biotite, is widespread but is generally apparent only microscopically. At several places rocks of the pluton are cut by swarms of aplite dikes, shown somewhat diagrammatically on the geologic map (pl. 1), or by large, fairly irregular masses of aplite, some of which locally grade to medium- or coarse-grained pegmatite.

[1] In 1971, Hibbard published a paper (see references), including a geologic map, that covers virtually the same area as the work cited here as "Hibbard (1962)". The report includes a substantial amount of new data and the map has been slightly expanded and substantially revised. Our paper was in press at the time, and unfortunately, we were unable to take advantage of this new contribution to the local geology in the present paper.

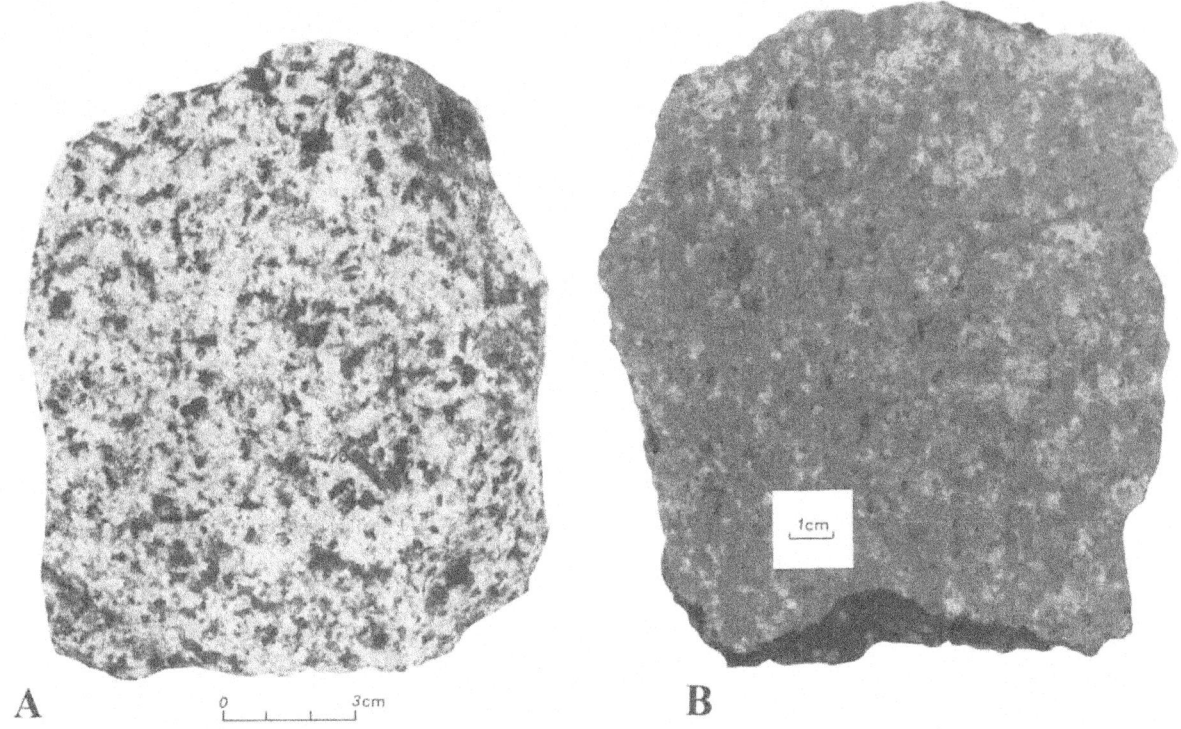

FIGURE 14.—Granodiorite of the Loomis pluton. A, typical hand specimen. B, sawed and stained slab of same specimen—plagioclase, medium gray; quartz and micro-cline, light gray; biotite and hornblende, black.

A small mass of trachytoid syenite cuts grano-diorite a mile and a half east of Gold Hill and is in-cluded with the Loomis pluton for convenience only; how it fits into the local sequence of batholitic intru-sion is not known. The closest lithologic equivalents are the monzonitic rocks within the Similkameen composite pluton.

Contacts

Contacts of the Loomis pluton with metamorphic wall rocks are sharp and the granitic rocks show only minor local evidence of contamination. At two locali-ties along Chopaka Creek, and at the north end of Aeneas Mountain, intrusion breccia has formed along the contact in amounts large enough to delineate on the map. Interfingering of granitic and metamorphic rocks is shown, somewhat diagrammatically, east of Sinlahekin Creek where the contact traverses the Sin-lahekin Game Range. There, the contact zone is marked by numerous apophysal dikes, local zones of

abundant inclusions, and detailed interleaving of gra-nitic and metamorphic rock over an area of several thousand square yards.

Contacts with other granitic rocks—Anderson Creek and Toats Coulee plutons—are generally less distinct, but it is possible to determine relative ages at a few localities. Although contacts between the Loomis and Anderson Creek plutons are sharp, relative ages are difficult to establish. At a locality near Bowers Lake, granodiorite of the Anderson Creek plu-ton clearly cuts a zone of mafic, hybrid-appearing breccia which grades to fine-grained diorite intrusive into quartz diorite of the Loomis pluton. At exposures near Olie Pass and east of the junction of Olie and Toats Coulee Creeks, inclusions of quartz diorite, identical with quartz diorite of the Loomis pluton, are enclosed by granodiorite of the Toats Coulee plu-ton. Diorite, present at both localities, appears to be of two ages, for one type is enclosed by the Loomis pluton, whereas another, somewhat finer grained va-riety, intrudes the Loomis.

Compositional Zoning

In order to determine whether or not the pluton shows evidence of zoning, three parameters—K-feldspar as percent of total feldspar, percent quartz, percent mafic minerals—were measured for 87 samples, and a fourth parameter, specific gravity, was measured for 113 samples. Of the four parameters, the relative abundance of K-feldspar, shown in figure 13, provides clear evidence of concentric zoning in the pluton. The contoured values show two north-trending highs in the southern two-thirds of the pluton, and another high of equal magnitude in the northern third, although the latter is cut out on the west by the Toats Coulee pluton. The abundance of quartz shows a correlative though somewhat less well-defined distribution pattern (fig. 11).

Little suggestion of zoning is shown by the contoured values of either specific gravity or the mafic mineral content (figs. 10 and 12). The density of data points is probably not sufficient to attribute much significance to local highs and lows shown by the contours. It is, however, worth noting that comparison between the sets of contoured data shows, not surprisingly, that some similarities exist between specific gravity and percentage of mafic minerals and between percentage of quartz and the relative abundance of K-feldspar.

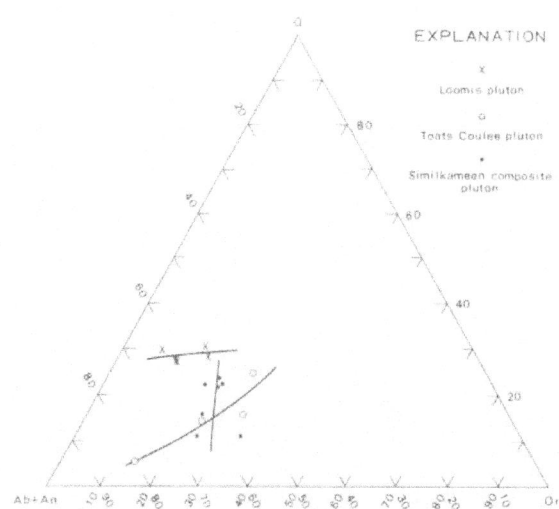

FIGURE 15.— Plots of normative quartz, orthoclase, and albite plus anorthite of quartz-bearing granitic rocks of the Loomis quadrangle.

Petrochemistry

To further investigate the composition of the pluton, a group of five samples, selected to represent the compositional range, were analyzed chemically for major constituents and spectrographically for minor elements; the results are recorded in table 3. Modes of the samples analyzed are plotted on figure 9, the norms on figure 15.

Major-element variation, graphed on a conventional Harker diagram in figure 16, shows that the Loomis pluton is strikingly distinct from the other plutons in its ratio of K_2O to silica. The Loomis also has a much higher alkali-lime index (Peacock, 1931), as

shown by figure 17, and, in addition, a trend line drawn by inspection through a plot of normative quartz, orthoclase, and albite plus anorthite (fig. 15) diverges sharply from that of the other plutons. The distinctive nature of the pluton is further expressed by the minor-element content, specifically by the total absence of Be, La, Nb, and by relatively low values of Sr.

Age

Potassium-argon ages have been determined for coexisting hornblende and biotite as 194 ± 6 and 179 ± 5 m.y., respectively, from a sample of the Loomis pluton collected along the Toats Coulee road close to BM 1490, in section 34, T. 39 N., R. 25 E. The determinations were made by Joan C. Engels of the U.S. Geological Survey. Her report with supporting analytical data is quoted at the end of the paragraph on the Toats Coulee pluton. Regarding the hornblende age (194 m.y.) as a minimum, but close to the accurate age for the Loomis pluton, the Loomis pluton is here considered Late Triassic.

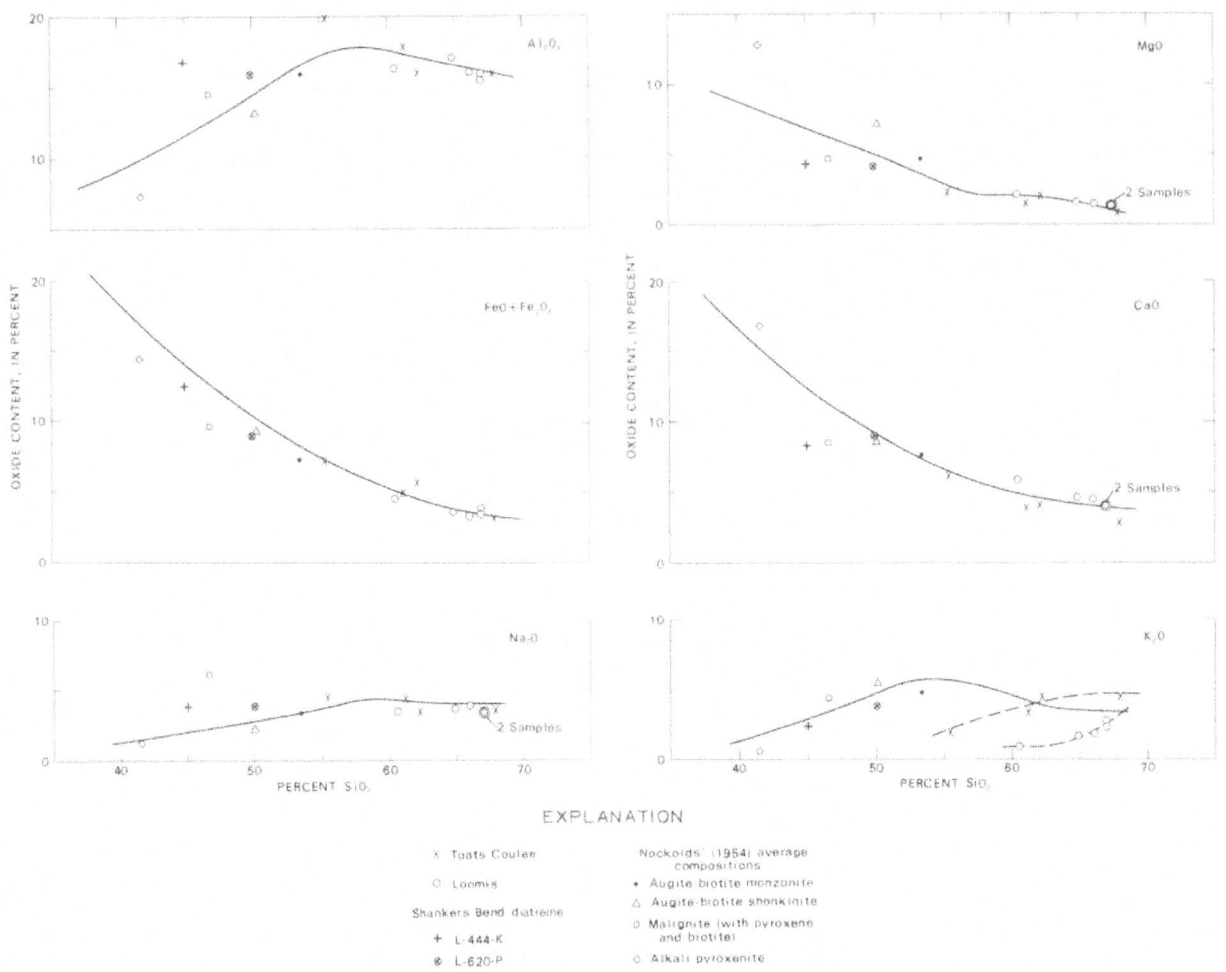

FIGURE 16.—Silica variation diagram comparing rocks from the Loomis and Toats Coulee plutons and the Shankers Bend diatreme, with the Similkameen composite pluton (solid line) and some average rocks of Nockolds (1954).

ANDERSON CREEK PLUTON

The Anderson Creek pluton is made up of five discrete, structureless granodiorite masses that intrude the Loomis pluton, Palmer Mountain Greenstone, and Kobau Formation along the northeast margin of the Loomis pluton; they are probably mutually contiguous beneath the alluvium of the Similkameen Valley. The name was taken from the creek that traverses the northernmost mass and was first used by Hibbard (1962), who applied it to the two northernmost masses. He correlated the two masses north and east of Chopaka Lake with the Toats Coulee pluton and the southernmost mass with the Loomis pluton. Hibbard's map

indicates that the pluton occupies less than a square mile west of the Loomis quadrangle boundary.

The pluton consists mostly of granodiorite as the modes in figure 9 indicate. The mafic mineral content averages about 10 percent and consists of hornblende, biotite, chlorite, and opaque minerals. Sphene is a conspicuous accessory mineral. The K-feldspar is typically microperthitic microcline but untwinned nonperthitic orthoclase is present locally. Plagioclase composition is generally in the An_{30-40} range but in one specimen the anorthite content is as low as An_{15}. The rock is typically medium grained but includes local fine-grained phases, and, as a whole, is finer grained than the other major plutons. The texture is

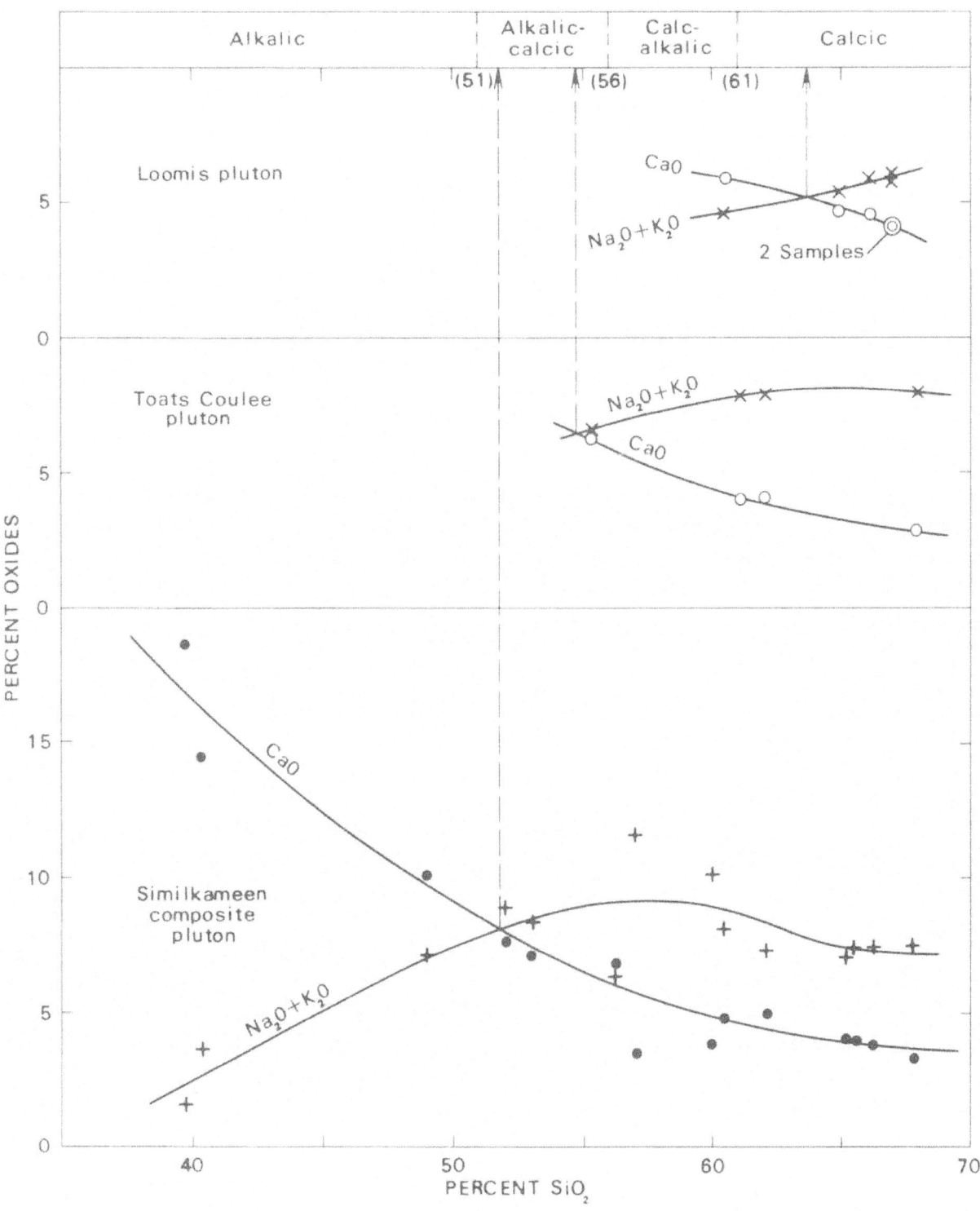

FIGURE 17.— Silica variation diagram comparing the alkali-lime indices (Peacock, 1931) of the Similkameen, Toats Coulee, and Loomis plutons.

hypautomorphic-granular except that weak porphyritic texture was noted in both the northernmost and southernmost masses. Moderate cataclasis has modified the texture locally. Although varied in texture and composition throughout, the two masses east of Bowers and Chopaka Lakes are the most mafic.

Contacts with metamorphic wall rock are typically steep and sharp; wall rock inclusions are sparse except near the southern end, where a sizeable exposure of intrusive breccia was noted. Contacts with the Loomis pluton are commonly sharp and are marked by zones of fine-grained diorite.

The Loomis pluton becomes increasingly gneissic adjacent to the contact but remains homogeneous and medium grained. The fine-grained diorite, on the other hand, is varied in texture and in color index at the contact but is not gneissic. Locally it contains inclusions of medium-grained quartz-bearing gneissic granitic rock resembling rocks of the Loomis pluton. These features suggest that the fine-grained diorite intruded the Loomis pluton. At one locality, near Bowers Lake, fine-grained diorite is extensively intruded by dikes of a coarser grained, more leucocratic rock, even approaching locally an intrusion breccia, which in turn is cut by granodiorite typical of the Anderson Creek pluton. Elsewhere the contact of the fine-grained diorite with rocks of the Anderson Creek pluton appears to be gradational over several tens of feet. The fine-grained diorite thus appears to be an early phase of the Anderson Creek intrusion.

The border-phase relations of the pluton are similar to those of the Toats Coulee pluton, and, in field exposures, Anderson Creek plutonic rocks locally bear textural resemblance to Toats Coulee plutonic rocks. These features, in addition to evidence that the Loomis pluton is intruded by the Anderson Creek pluton, form the basis for assigning a Jurrasic age to the Anderson Creek pluton.

TOATS COULEE PLUTON

Within the Loomis quadrangle, rocks of the Toats Coulee pluton range in composition from diorite to quartz monzonite but have an average composition of granodiorite. The name, first used by Hibbard (1962), is taken from the creek that bisects the exposure of the pluton in the Loomis quadrangle. Hibbard's map shows that the pluton occupies an area of at least 70 square miles west of the Loomis quadrangle.

In addition to minerals listed in table 3, relict augite was recognized in a few thin sections; and zircon, enclosed by biotite, was seen in nearly all thin sections examined. The rocks are characterized by marked variation in grain size, texture, and composition; the extremes locally occurring within a distance of a few tens of feet of each other, particularly near the contact with the Loomis pluton. The grain size ranges from fine to medium; the texture is commonly hypautomorphic-granular (fig. 18) but is locally diabasic, as in some of the mafic border rocks, or porphyritic (fig. 19) with subsequent locally poikilitic microcline phenocrysts which attain maximum dimensions of 2 cm.

Contacts and Border Phase

The fine-grained dioritic border phase of the Toats Coulee pluton, in part separately delineated on the geologic map (pl. 1), extends southwest along the contact to the Loomis quadrangle boundary but was not mapped separately there. Its chief characteristic is variation in both composition and texture, which generally appears random but locally is sufficiently regular to define a gently dipping layered structure that is probably best exposed on the southern slopes of Quartz Mountain. Gradation of the border phase into granodiorite typical of the remainder of the pluton is the basis for including it as part of the Toats Coulee pluton. In several places, however, particularly north of Quartz Mountain, granodiorite of the pluton clearly cuts the mafic border phase, locally producing intrusion breccias and indicating that the border phase is, at least in part, somewhat older. It is significant that whereas the typical granitic rocks of the pluton commonly intergrade with dioritic rock of the border phase, no comparable gradation was noted between the border phase dioritic rock and quartz diorite of the Loomis pluton

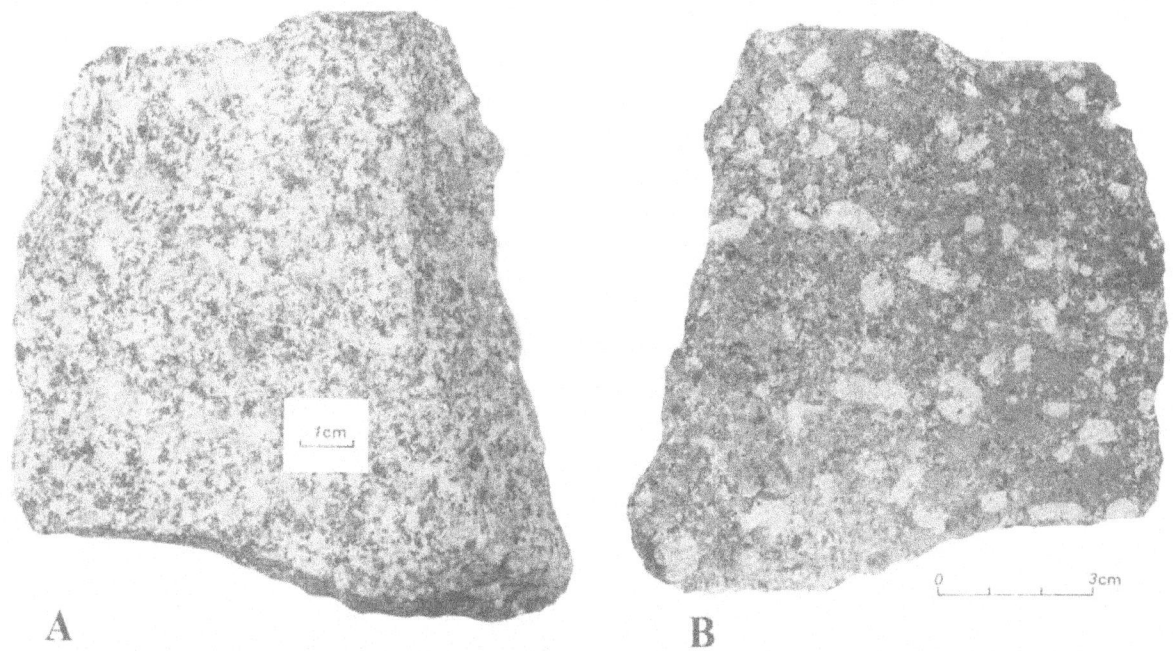

FIGURE 18.—Granodiorite of the Toats Coulee pluton. A, typical hand specimen. B, sawed and stained slab of same specimen—plagioclase, dark gray; microcline and quartz, light gray; biotite and hornblende, black.

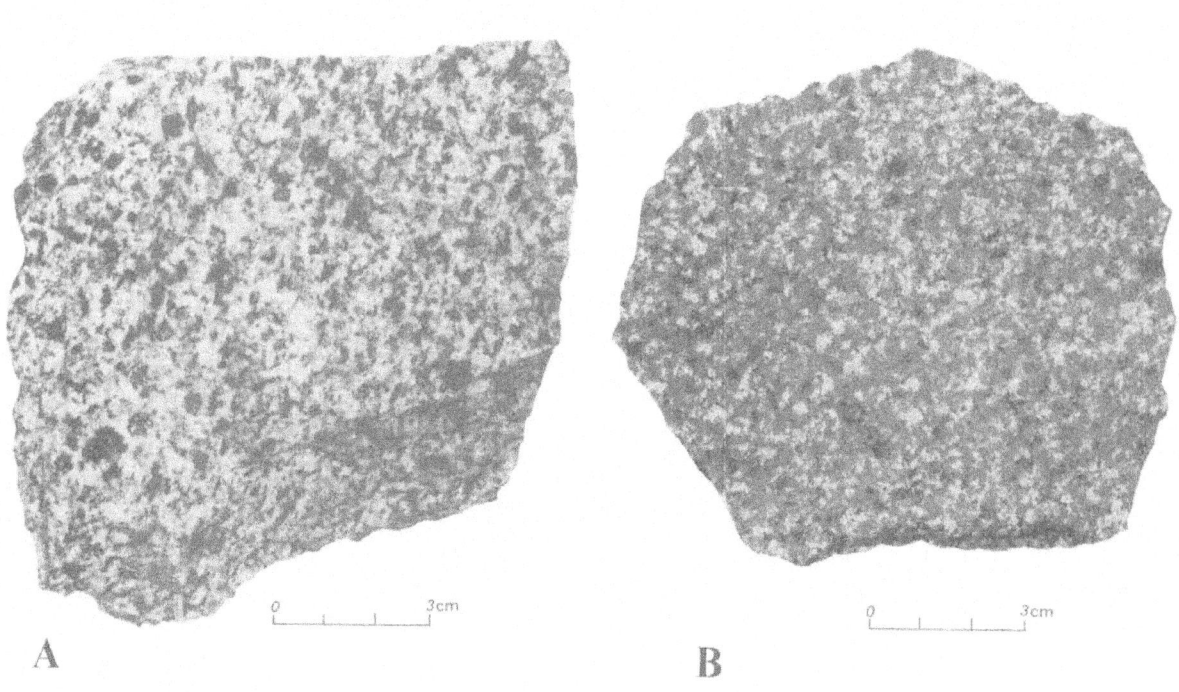

FIGURE 19.—Porphyritic quartz monzonite of the Toats Coulee pluton. A, typical hand specimen. B, sawed and stained slab of same specimen—plagioclase, dark gray; quartz and microcline phenocrysts, light gray; biotite and hornblende, black.

adjacent to the contact. Contacts with the Loomis are typically sharp. This is compatible with the evidence that rocks of the Loomis pluton are locally intruded both by dioritic rock and by more felsic rocks of the Toats Coulee pluton. Coarser grained inclusions of diorite, observed in the Loomis pluton near contacts with the Toats Coulee, presumably represent remnants of a pre-Toats Coulee wall rock rather than the diorite phase of the Toats Coulee.

Petrochemistry

To investigate the chemical composition of the Toats Coulee pluton, four selected samples representing its compositional range were analyzed chemically and spectrographically; the results are recorded on table 3. The Toats Coulee pluton is chemically distinct from the Loomis and Similkameen plutons, as shown on figures 15, 16, and 17. Its singularity in quantitative mineralogy is shown by the trend of the modes shown on figure 9. A related distinction is shown by the alkali-lime index on figure 17. A subtle difference in both Na_2O and CaO may also possibly exist, but the validity of this difference is dubious when the plotted points for these oxides are compared with the points (fig. 20) from which the curves of figure 16 were drawn.

In virtually all of the comparisons referred to above, the Toats Coulee occupies a position intermediate between the Loomis and the Similkameen plutons; it is markedly more alkalic than the Loomis but less alkalic than the Similkameen.

Age

Potassium-argon ages have been determined for coexisting hornblende and biotite from a sample of the

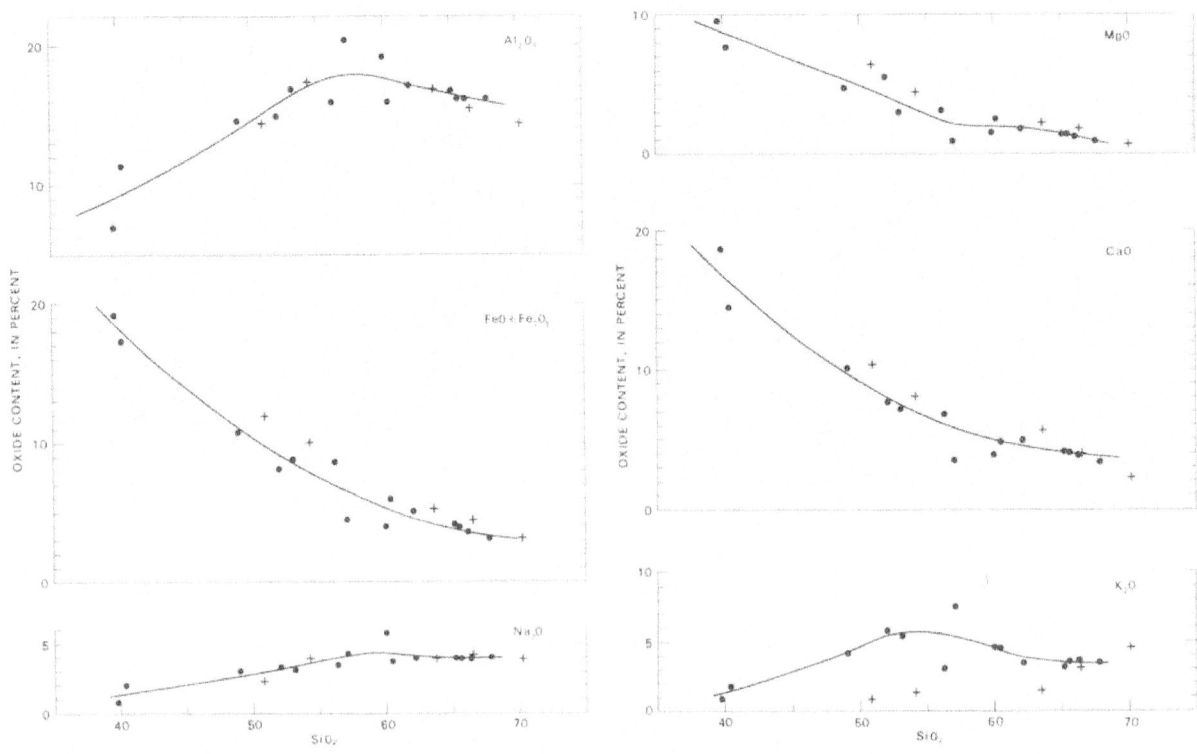

FIGURE 20.—Silica variation diagram of the Similkameen composite pluton. Shown for comparison, by plus signs, are Nockolds' (1954) average dellenite, rhyodacite, dacite, andesite, and normal tholeiitic basalt.

Toats Coulee pluton that was collected a mile west of the Loomis quadrangle boundary, northeast of the junction of Toats Coulee and Ninemile creeks, in the SW$\frac{1}{4}$ sec. 13, T. 39 N., R. 24 E. The determinations were made by Joan C. Engels of the U.S. Geological Survey in Menlo Park, Calif. Her report on the age of the Toats Coulee pluton follows (written communication, 1967):

Loomis pluton cuts rocks of known Late(?) Permian age and also intrudes rocks that may be as young as Triassic. As stated earlier, the Loomis pluton is therefore considered to be of Late Triassic age. The Toats Coulee intrudes the Loomis and may itself be partly responsible for the discordant ages from the Loomis, although this, of course, does not explain the discordant mineral ages in the Toats Coulee. Viewing the hornblende age as a

Analytical Data

Mineral	K_2O (percent)	Ar^{40*} (moles/g)	Atmospheric Ar (percent)	Age (m.y.)
(Toats Coulee pluton, sample no. L-591)				
Biotite (replicate)	8.80 8.95	2.057×10^{-9} 2.064×10^{-9}	3.9 3.0	151 ± 5
Hornblende (replicate)	1.25 1.28	3.288×10^{-10} 3.347×10^{-10}	7.4 6.5	170 ± 5
(Loomis pluton, sample no. L-498-A)				
Biotite (replicate, K_2O only)	8.74 8.74	2.421×10^{-9}	33.4	179 ± 5
Hornblende (replicate, K_2O only)	0.627 0.626	1.886×10^{-10}	5.3	194 ± 6

* Radiogenic

The biotite-hornblende pairs show a discordance that is outside the limits of analytical error. This is confirmed by good agreement on replicate determinations of samples from the Toats Coulee pluton. Such a discordance could be due to some later event of wide areal extent nearby, or to the effects of chloritization and weathering, either of which would tend to lower the apparent age of biotite relative to that of hornblende.

Miss Engels believes, as we do, that the hornblende age in both cases closely approximates the true ages of the plutons, because, on geologic grounds, the

minimum, but close to the true age, the age of the Toats Coulee pluton is therefore considered to be Early Jurassic.

SIMILKAMEEN COMPOSITE PLUTON

The Similkameen composite pluton is a zoned pluton that is quartz monzonite and granodiorite in the central part, a complex of alkalic rocks in the marginal part, and monzonite in the intervening part.

As defined here, the composite pluton comprises mapped bodies of granodiorite, pyroxenite, malignite, syenitic gneiss, a small body of greisen, and one of alaskite. Thus defined, it combines Daly's (1906, 1912) "Similkameen batholith" and his "Kruger alkaline body" as one genetically related unit. This is in accord with his conclusions that both the Similkameen and the alkalic rocks ". . . belong to one petrogenic cycle" (Daly, 1912, p. 459). Mapping to the north (Bostock, 1940) and to the west (Hibbard, 1962) shows the Similkameen to be oblong in plan, trending slightly north of west and occupying a total area of about 130 square miles (pl. 1, inset). About 20 percent of the pluton is within the Loomis quadrangle.

Quartz Monzonite—Granodiorite

Quantitatively, granitic rocks compose more than 90 percent of the Similkameen pluton, and they range in composition from monzonite through quartz monzonite to granodiorite. Judging by the modal plots in figure 9, the average composition lies near the boundary between quartz monzonite and granodiorite. The modes of the analyzed specimens, listed on table 4, are representative of the part of the pluton within the Loomis quadrangle. The ubiquitous occurrence of epidote as a megascopically visible mineral is a conspicuous and characteristic feature of the pluton; thin sections show that epidote is invariably associated with, and probably altered from, the ferromagnesian minerals. Sphene is a distinctive, megascopically visible accessory. Microscopic examination shows that relict pyroxene crystals, largely replaced by amphibole, are fairly common. Examination in ultraviolet light showed zircon to be present in all samples. The rock is sporadically porphyritic (fig. 21) with scattered tabular, subhedral to euhedral, poikilitic to homogeneous phenocrysts of microperthitic microcline that contrast markedly with the ragged, poikilitic microcline phenocrysts sparsely distributed in the Loomis pluton, or the larger blocky ones in the Whisky Mountain pluton. Xenomorphic granular texture is prevalent in the Similkameen but uncommon in rocks of the other plutons.

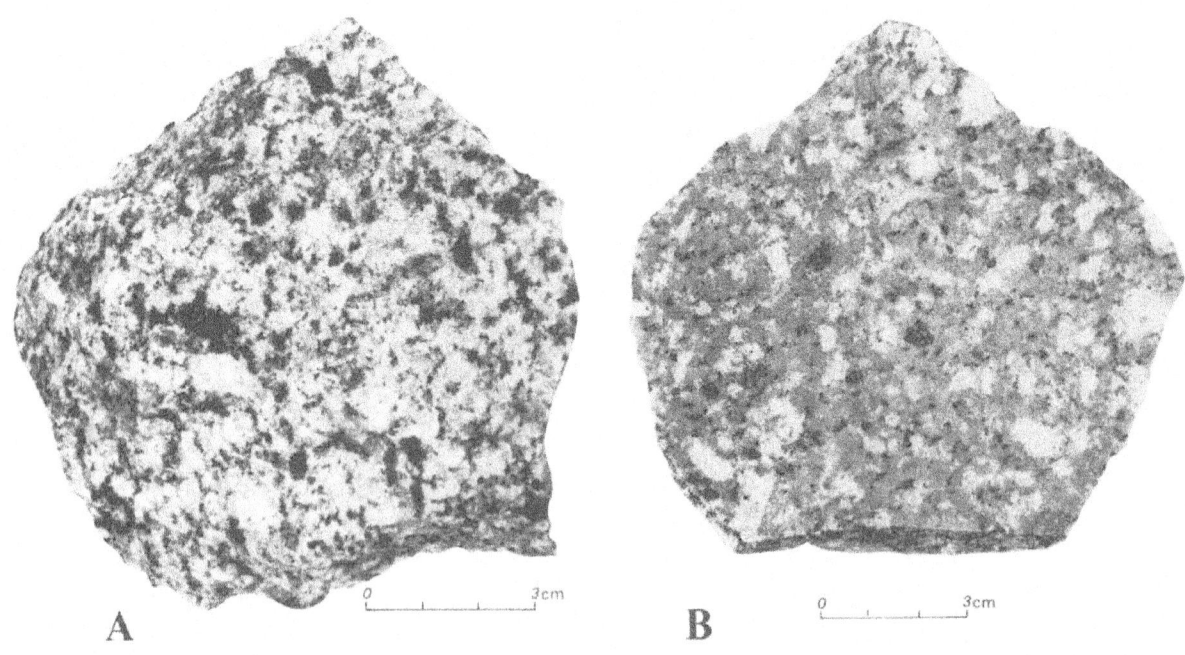

A

B

FIGURE 21.—Granodiorite of the Similkameen composite pluton. A, typical hand specimen. B, sawed and stained slab of same specimen—plagioclase, dark gray; quartz and microcline phenocrysts, light gray; biotite and hornblende, black.

Pyroxenite, Malignite, and
Syenite Gneiss

Along more than half of its exposed length in the Loomis quadrangle, the border of the Similkameen composite pluton is a complex of pyroxenite, malignite, and syenite gneiss, all of which intergrade and are also gradational with the granitic rocks. These units are contiguous with the unit designated "Kruger syenite" by Bostock (1940). Leucocratic diaschistic dikes with sharp contacts locally intrude both malignite and pyroxenite; contacts with wall rocks are typically concordant and sharp. Field identification of the border rocks was based largely on relative abundance of mafic constituents, although accompanying structural and textural features were also of importance. Quartz is sparse or absent in all the border rocks, and the adjacent granitic rock is likewise quartz-poor but shows a progressive increase in quartz away from the border, toward the center of the pluton (fig. 11). This corroborates Bostock's findings north of the quadrangle where he recognized "a concentrically zoned structure centred where Similkameen River crosses the International Boundary" (Bostock, 1940). In all of the border rocks the common mafic minerals are augite, hastingsite, and biotite; apatite and garnet are fairly common accessories and sphene and epidote are generally present and megascopically visible. The syenite gneiss is commonly leucocratic but its composition is varied and the unit ranges from syenite to shonkinite. The analyzed specimen (table 4) is shonkinite but the term "syenite" is retained for the unit as a whole to emphasize its general leucocratic nature. The gneissosity is defined by alternating layers of contrasting abundances of mafic minerals; dark layers are rich in biotite, magnetite, epidote, and locally garnet, whereas light layers are rich in feldspars and muscovite. Layers range in thickness from a few millimeters to a few centimeters. Microscopically, the gneiss shows a weak directional fabric, defined chiefly by preferred orientation of subhedral biotite. The feldspars also tend to be subhedral and elongate in the plane of foliation. Muscovite, conversely, is anhedral, showing no parallelism between physical shape and foliation

but showing some tendency toward parallelism between foliation and the basal cleavage.

In the field the presence of feldspar in rocks of the malignite unit served to distinguish them from pyroxenite, in which feldspar is virtually absent; microscopic examination shows little qualitative difference in mineralogy. Contacts between the two rocks are gradational over tens or hundreds of feet nearly everywhere, but in a few places dikes of malignite (fig. 22) cut pyroxenite, showing that the malignite is younger, at least in part. Textures of the rocks are commonly xenomorphic-granular (fig. 23), but the malignite also commonly displays both porphyritic and trachytoid texture, the latter shown by conspicuous euhedral, tabular K-feldspar phenocrysts, in at least moderate alignment, set in a xenomorphic-granular matrix (fig. 22). Relative abundances of hastingsite, augite, and biotite range widely in both rocks, though pyroxene is typically the dominant mineral in the pyroxenite. The minimum color index of the malignite is 30 and proportions of hastingsite-augite-biotite commonly range from 1:1:1 to 3:1: <<1. Nepheline or its alteration products were rarely found in more than accessory amounts in samples of the malignite collected from south of the International Boundary. Daly (1912, p. 451) reports 5.4 percent nepheline in an analyzed sample, and Campbell (1939, p. 538) gives 13 percent nepheline as the average of 8 Rosiwall counts. These data suggest a southward decrease in abundance of feldspathoids within the malignite unit, and the rocks of the small part of the unit within the Loomis quadrangle probably would be more properly classified as shonkinite.

Greisen, Aplite, and Alaskite

A small patch of greisen is present near American Butte, which is near the northern boundary of the Loomis quadrangle. It is composed of muscovite, quartz, albite, chlorite, calcite, and small amounts of pyrite, and is gradational with the adjacent rock. Dikes of aplite and alaskite occur sporadically throughout the Similkameen composite pluton and one alaskite body, along the southeast boundary of the pluton, is

FIGURE 22.—Specimen of trachytoid malignite from dike cutting pyroxenite. Contains pyroxenite inclusion. Shows extreme development of tabular microcline phenocrysts.

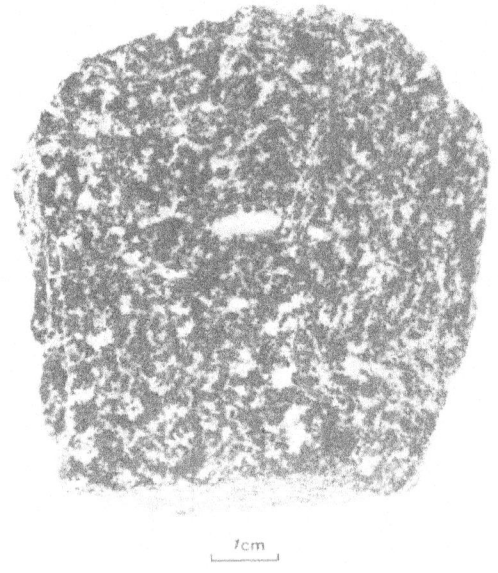

FIGURE 23 — Specimen of shonkinite showing typical xenomorphic-granular texture. Hastingsite, augite, and biotite, black; microcline, a small amount of plagioclase, and nepheline (rare), light gray.

large enough to be shown on the geologic map, plate 1. It consists chiefly of quartz and microcline and is clearly younger than the enclosing rocks.

Petrographic Trends

To investigate the distribution of several compositional properties, the following values from a suite of modally analyzed samples of the leucocratic rocks were plotted on the map and contoured: specific gravity, percentage of quartz, percentage of mafic minerals, and ratio of microcline to total feldspar (see figs. 10, 11, 12, 13). Figures 10, 11, and 12 show an irregular but unmistakable tendency toward concentric zoning with respect to the outline of the pluton, the quartz content bearing an expected, generally reciprocal relation to specific gravity and mafic mineral content, although detailed correlation among the three parameters is only fair. The pattern of the microcline to total feldspar contours does not follow that of the other parameters. Instead, it shows more irregularities and reversals in trends inward from the margin of the pluton; but, despite this, the pattern roughly parallels the pluton contacts. All four parameters show the greatest irregularities in their contoured distributions near the eastern and southeastern margins of the pluton. Significantly, these are the margins where the mafic alkalic phases are most abundant. Irregularities not-withstanding, there is little doubt that the one-fifth of the Similkameen composite pluton that occurs within the Loomis quadrangle and that was studied in the present investigation shows definite, though somewhat irregular, concentric compositional zoning.

Petrochemistry

The chemical composition of the Similkameen composite pluton was examined by obtaining chemical and spectrographic analyses of 14 samples representative of much of the compositional range of the pluton. The analyses, modes, and norms are listed in table 4.

The oxides as plotted on the silica variation diagram of figure 20 show serial variation with moderate scatter. Plotted for comparison on the same diagram are some average calc-alkalic rocks of Nockolds (1954), which show generally parallel, nearly coincident trends of all oxides except K_2O. Except for rocks with the highest silica content, the Similkameen composite pluton is markedly richer in K_2O than the average rocks of Nockolds. This alkalic character is perhaps more strikingly shown on figure 17, where an alkali-lime index of less than 52 classifies the pluton as alkali-calcic according to the scheme of Peacock (1931).

Although rocks of the Similkameen that are intermediate in composition between the pyroxenite and granodiorite were classified as malignite by Daly (1912), Campbell (1939), and Lounsbury (1951), figure 16 shows that Nockolds' (1954) average monzonite and shonkinite plot closer to the Similkameen curves than his malignite. This is not surprising, however, since our sample collections yielded specimens poorer in nepheline than some collected by the earlier workers, and thus are more akin to shonkinite than malignite. Continued use of the name "malignite" for this intermediate group is justified, however, because (1) no more appropriate name exists for the group as a whole; (2) the local malignite unit surely includes rocks corresponding to Lawson's (1896) originally defined malignite; (3) literature references

to the local rocks as malignites extend back through more than 50 years.

Origin and Age

On the nature of field relations between the various rock types here included as part of the Similkameen composite pluton, there is general accord among all those who have studied the terrane— Daly (1912), Campbell (1939), Bostock (1940), Lounsbury (1951), and ourselves—not only that mutual intergradation exists among all of the principal rock types, but also that dikes and local crosscutting relations indicate a direct relation between increasing age and increasing mafic content. Hence, it is not surprising that the conclusions based on these relations are accordingly similar: the composite pluton, although representing one intrusive episode, shows a mafic-to-felsic trend with time.

Little field evidence of the relative age of the Similkameen was seen other than that it intrudes both the Kobau Formation and a serpentinite mass, which is intrusive into the Kobau on Little Chopaka Mountain. Indirect evidence of its age, considered to be Jurassic or Cretaceous, is discussed in the sections on metamorphism, Shankers Bend diatreme, and Ellemeham Formation.

PETROLOGIC SUMMARY OF THE PLUTONIC ROCKS

The modal composition of the granitic rocks in the Loomis quadrangle, summarized on triangular diagrams (fig. 9), indicates that the three major plutons—Loomis, Toats Coulee, and Similkameen— are compositionally distinct. Median lines drawn through the modes show striking differences in the compositional trend of each pluton, doubtless reflecting fundamental differences in their magmatic and intrusive history. Median lines drawn through the norms of these same plutons (fig. 15) show trends that lie parallel and close to the modal trends. No trends were recognized

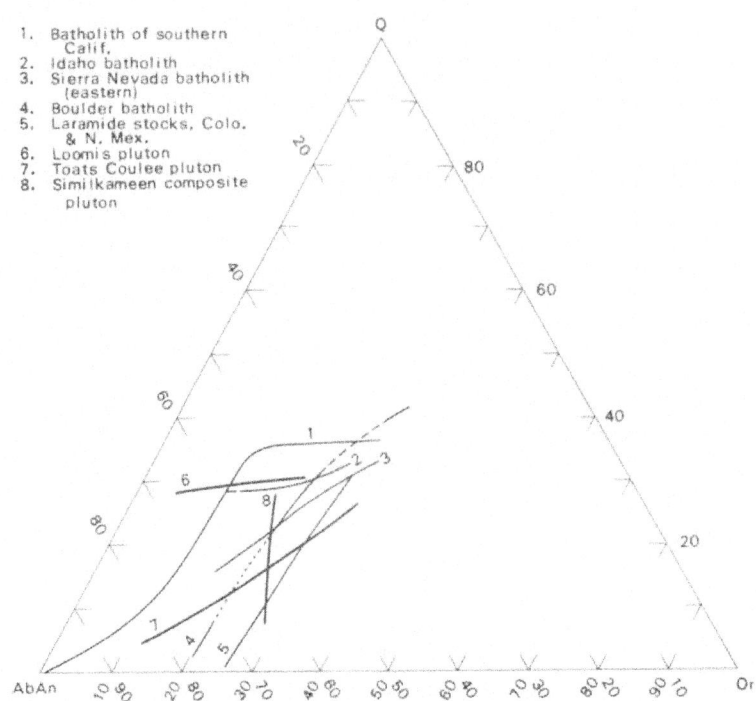

1. Batholith of southern
   Calif.
2. Idaho batholith
3. Sierra Nevada batholith
   (eastern)
4. Boulder batholith
5. Laramide stocks, Colo.
   & N. Mex.
6. Loomis pluton
7. Toats Coulee pluton
8. Similkameen composite
   pluton

FIGURE 24.—Normative trends of three plutons in the Loomis quadrangle, compared with
normative trends of some batholiths in the western United States. (After Bateman
and others, 1963.)

in either the Whisky Mountain or Anderson Creek plutons.

Conclusions based on the foregoing observations are: (1) Magmas from which each of the three major plutons crystallized were distinctly different, chiefly in K_2O content. (2) Differentiation trends of the magmas were mutually distinct, possibly in part a result of differing K_2O contents. (3) If the Similkameen is younger than the Toats Coulee, as is suspected, the compositions of the major plutons suggest a trend toward increasingly alkalic magmas with time.

A comparison of the normative trends with those constructed for some well-known granitic suites (Bateman and others, 1963) is shown on figure 24, where it can be seen that the only trends approaching coincidence are those of the Idaho batholith and the Loomis pluton. The Toats Coulee trend is parallel to that of the eastern part of the Sierra Nevada batholith but, for any feldspar ratio, the Toats Coulee is about 6 percent poorer in normative quartz. The Similkameen trend has no parallel

counterpart on the diagram but its steep angle with respect to the feldspar side of the triangle is a feature most closely shared by the trends of the Boulder batholith and the Laramide stocks of Colorado and New Mexico. The latter are rocks with known alkalic affinities, as is the Similkameen composite pluton, whereas the Idaho and southern California batholiths are both known to be strongly calc-alkalic, as is the Loomis pluton. The Toats Coulee pluton and the plutonic rocks of the eastern Sierra Nevada both are intermediate in their alkalic-calc-alkalic character. It should also be noted that in the comparisons made above, individual plutons in the Loomis quadrangle are compared with linear trends defined by groups of genetically related plutons from other batholithic areas. It is obvious from inspection of figure 15, however, that as a group the norms of the three local plutons define no recognizable trend, differing in this respect with many batholithic terranes. The lack of a collective trend is also manifested by the wide range in K_2O values shown on figure 16.

SHANKERS BEND DIATREME

The Shankers Bend diatreme occupies an area of less than a square mile that is surmounted by two low hills, about a mile south of Shankers Bend near the northeast corner of the quadrangle. The diatreme is an irregular, asymmetrical, concentrically zoned structure whose core is occupied by a complex of ma-lignite, fenite (syenitic product of alkalic metasomatism accompanied by partial mobilization and partial de-silication), and related alkalic rocks. The rocks of the core grade outward to a zone of fenitized green-stone and fenitized greenstone breccia (fig. 25) of

0   3cm

FIGURE 25.—Fenitized greenstone breccia from Shankers Bend diatreme. Angular dark fragments are mafic hornfels (after green-stone?) enclosed in a weakly gneissic syenitic matrix composed chiefly of micro-cline (note light-colored porphyroblasts), oligoclase, hastingsite, and biotite.

the Ellemeham and possibly Kobau Formations, which in turn grades to an outer zone where unreplaced rocks of both formations are shattered and brecciated. The unique distribution and structure of the outer zones suggest that the rocks of the core were, at least in part, explosively injected into their host.

The most striking effects on the rocks surrounding the core are extreme brecciation and extensive feniti-zation; the latter process has created analcitic, shon-kinitic, and malignitic gneisses from greenstones of the Ellemeham and possibly of the Kobau as well.

A zone of metamorphism, generally a few tens of feet wide, irregularly surrounds the fenite gneiss and malignite of the core; it can be recognized by visi-ble increase in the size of mica flakes. The contact itself is highly irregular, and along it, all stages in the conversion of greenstone to fenite can be seen, including local development of medium-grained trachy-toid texture.

Brecciation is most dramatically shown in green-stone of the Ellemeham Formation on the southwestern flank of the westernmost of the two hills. From the base to the top of the hill, layered, gently folded greenstone becomes increasingly shattered until, near the crest, it is a chaotic jumble of fragments. In nu-merous outcrops, however, bedding can be traced for a few feet, indicating that differential movement of fragments has been slight. Much of the fenite in the area also has been shattered and brecciated.

Many discontinuous dikes of monzonitic alloclas-tic intrusive breccia (Wright and Bowes, 1963, p. 83), a few inches to a few tens of feet thick, cut the brec-ciated and nonbrecciated fenite and are most abundant on the easternmost hill. Strikes are northeast or north-west with moderate to steep dips northward. The dikes pinch and swell, anastomose locally, and range in length from a few feet to as much as a few hundred feet. They are composed chiefly of angular fragments of xenomorphic-granular feldspathic rock that is locally gneissic and less mafic than the host; the fragments grade in size serially from a maximum of a few inches to submicroscopic and do not appear to be recrystal-lized. Less abundant are carbonate masses that resemble the breccia dikes in habit and dimensions. Some, but not all, are obviously breccias. A few contain signifi-cant amounts of fragmented feldspar crystals and one contains accessory celsian. The carbonate masses are believed to be carbonatite dikes because of their many physical similarities to the feldspathic dikes and their association with the alkalic rocks of the diatreme.

Fenite breccia is overlain by Tertiary conglom-erate on the eastern flank of the easternmost hill, and because the conglomerate consists almost entirely of granitoid detritus, including fenite, its contact with the fenite breccia is difficult to locate precisely. Reworking of the shattered rock of the Ellemeham to form conglomerate is evident in exposures scattered over the crest of the western hill.

The composition of the fenitized rocks is not unique in the area, because similar rocks form the syenitic and malignitic border of the Similkameen composite pluton. There, however, the country rock is not so obviously fenitized and its contact with gneiss and malignite is typically sharp. Furthermore, breccia dikes or brecciated country rock are not features of the alkalic border zone of the Similkameen. Nevertheless, the striking similarity between the rocks of these two terranes in composition and character and their fairly close proximity suggest that they were probably emplaced during the same episode of plutonism and are derived from the same magmatic source.

Analyses of a shonkinite and a fenitized greenstone from the diatreme are listed in table 5, along with an analysis of a nonfenitized greenstone that was collected several miles south of the diatreme. Both the mode and norm of the shonkinite from the diatreme closely resemble those of the shonkinite of the Similkameen composite pluton (table 4), and the similarity extends to the texture, both megascopic and microscopic. On the silica-variation diagram of figure 18, the shonkinite shows fair agreement with the curves of the Similkameen suite, but the fenitized greenstone plots considerably off most of the curves. The latter unmistakably shows the effects of potassium metasomatism, however, and its K_2O value lies close to the Similkameen K_2O curve. There is no appreciable difference in minor-element content between the fenite and malignite, nor between them and the minor-element content of the Similkameen suite. The nonfenitized greenstone has been altered somewhat, judging by its relatively high H_2O+ and CO_2 content, but it compares reasonably well with Nockolds' average tholeiitic basalt. The fenite was probably derived from a similar rock.

INTRUSIVE ROCKS IN DIKES AND SMALL MASSES

Dikes and small masses of intrusive rocks of varied compositions, textures, and dimensions abound in both plutonic and metamorphic terranes of the Loomis quadrangle, but only the largest are shown on the geologic map. They were not studied in detail and are classified below and designated on the map according to criteria by which they were distinguished in the field.

Intermediate to felsic rocks—medium to light colored, aphanitic, or porphyritic (feldspar phenocrysts) with aphanitic matrix. Specific gravity generally < 2.80. Dikes of this class are locally bleached, altered, and are hosts to sulfide mineralization.

Mafic rocks—dark colored, typically aphanitic. Specific gravity generally > 2.80.

Granitic rocks—typically medium grained, quartz dioritic to quartz monzonitic in composition.

Aplite and alaskite—fine to medium grained, locally grading to pegmatite; saccharoidal texture common; mafic minerals sparse or absent.

Volcanic rocks—medium to dark gray or brown; generally less dense than other dikes and often softer; porphyritic with aphanitic matrix.

ROCKS OF TERTIARY AGE

SEDIMENTARY ROCKS

Irregular, north-south-trending bluffs and cuestas along the northeastern border of the Loomis quadrangle mark the western extent of a thick pile of Tertiary conglomerate and arkose, which evidently accumulated in an intermontane basin centered several miles east of the quadrangle. The Tertiary rocks are little deformed and only locally faulted, although they have been warped so that within the quadrangle they present a homoclinal sequence tilted to the east at dips of about 30 degrees.

Aggregate thickness of the Tertiary clastic rocks is about 4,000 feet and they are informally divided into two formations: the lower, a thick sequence of conglomerate and arkose; and the upper, a much thinner conglomeratic graywacke containing clasts of volcanic rock.

Conglomerate and Arkose

The conglomerate and arkose sequence comprises three members, in ascending order: a basal conglomerate member composed predominantly of metamorphic detritus; a conglomerate member composed predominantly of granitic detritus; and an arkose member. The conglomerates locally total as much as 2,200 feet in thickness, and the arkose is at least 1,600 feet thick. The conglomerate members contain thin interbeds of arkose within the Loomis quadrangle, which thicken abruptly to the east along with concomitant thinning of the conglomerate. Ultimately the conglomerates wedge out and the arkosic interbeds are continuous with the arkose overlying the conglomerate in the Loomis quadrangle.

Conglomerate derived chiefly from metamorphic rocks.—This conglomerate, the basal member within the clastic section, is composed chiefly of rounded to subrounded fragments of metamorphic rock, ranging from less than an inch to as much as a few feet in maximum dimension but averaging in the 1- to 4-inch range, set in a graywacke matrix. The rock is structureless except for thin local lenses of thin- to thick-bedded arkose that are sparse in the Loomis quadrangle but become more numerous and thicken markedly to the east in the Oroville quadrangle. Except for the arkose lenses this rock is virtually indistinguishable from the upper member of the Ellemeham Formation, which it overlies. The arkose unit at the base of the metamorphic-bearing conglomerate thickens eastward and, in the Oroville quadrangle, interfingers with the overlying conglomerate and grades into the younger arkose member. The metamorphic-bearing conglomerate, which is thickest in the southwestern part of the Tertiary belt, thins rapidly and pinches out to the north and east.

Conglomerate derived chiefly from granitic rocks.—The upper of the two conglomerate members is highly distinctive because it is composed chiefly of cobbles of rock types peculiar to the Similkameen composite pluton, which lies to the northwest. The bulk of the cobbles are granodiorite, but malignite and trachytoid syenite are well represented. In addition, the conglomerate contains a sprinkling of cobbles of metachert and greenstone derived from the older formations. The clasts are oval to irregular in shape and as much as 4 feet in diameter, though most are in the 4- to 10-inch-size range, and are cemented by a greenish-gray very fine to coarse arkosic matrix. The granite-bearing conglomerate is a hard, resistant rock cropping out in rounded, convex slopes or light-gray ledgy cliffs as seen in the gorge of the Similkameen River. Although a gross bedded aspect is evident when the cliff exposures are viewed from a distance, bedding is rare except for sparse thin lenses of arkose. In the northernmost part of the Tertiary belt, the granite-bearing conglomerate member directly overlies the Kobau Formation and greenstone of the Ellemeham Formation with marked unconformity.

Arkose.—The arkose member is composed of interbedded porous, weakly indurated, light-olive-gray to yellowish-gray arkose, pebbly arkose, fine pebble conglomerate, and siltstone. Quartz, feldspar, and as much as 10 percent dark minerals are the chief components of the arkosic beds, usually present as fine to coarse subangular grains weakly cemented in a clay and carbonate matrix. Most of the arkosic beds contain scattered pebbles averaging half an inch in diameter, mainly of rounded fragments of metamorphic rock. Interbeds of granule conglomerate are more abundant in the upper part. The conglomerates are composed of rounded pebbles of metamorphic rock, commonly about half an inch in diameter but ranging up to cobbles 10 inches in diameter, set in an arkosic matrix. Thin light-olive-gray siltstone interbeds are common, usually grading downward to arkose and capped by a bed of more indurated arkose or conglomerate. Bedding is well defined and crossbeds and ripple marks are common.

Volcanic Graywacke

The volcanic graywacke is a heterogeneous assemblage of graywacke and conglomerate typified by the ubiquitous occurrence of clasts of aphanitic to fine-grained porphyritic volcanic rock. The graywacke is commonly soft, porous, and friable, and consequently outcrops are poor. Within the Loomis quadrangle the unit crops out as a ledgy slope in the fault block east

of Ellemeham Draw (NW¼ sec. 24, T. 40 N., R. 26 E.).
Conglomerate containing rounded cobbles and pebbles
of dense, grayish-red-purple, light-gray, and greenish-
gray aphanitic volcanic rock, locally with biotite and
(or) plagioclase phenocrysts, bonded in a matrix of
crystal tuff, grades upward into medium-grained, light-
olive-gray graywacke and granule conglomerate whose
clasts are predominantly angular fragments of heterogeneous metamorphic rock types with only a sprinkling
of volcanic clasts.

Elsewhere in the quadrangle the presence of the
volcanic graywacke unit is inferred beneath sandy col-
luvium containing abundant volcanic pebbles. Exposures are somewhat better to the east in the Oroville
quadrangle. There the unit contains, in places, a
conspicuous zone of pale-green to grayish-green crystal lithic tuff at or near the base. The tuff contains
broken to euhedral delicately zoned oligoclase-
andesine crystals, quartz crystals, intergrowths of
calcite-chlorite pseudomorphous after hornblende(?),
and clasts of devitrified glass in a matrix of weakly
birefringent fine-grained chlorite. The proportion of
volcanic constituents declines from the base upward,
and the lithic tuff grades to light-olive-gray graywacke
similar to that constituting the upper part of the unit in
the Loomis quadrangle.

The basal contact was not observed, but field
relations suggest that it is marked by a slight angular
unconformity. A similar sequence of Tertiary strata at
Whitestone Mountain, 12 miles to the south in the
Oroville quadrangle, is provisionally correlated with
the units herein described. There the volcanic rock-
bearing graywacke overlies the arkose member of the
arkose and conglomerate formation along a very uneven
surface, although bedding in the two units is grossly
parallel.

Age

Tertiary sedimentary rocks are cut by the hornblende-bearing dacite plugs dated as early Eocene by
potassium-argon analysis and thus must themselves be
Eocene or older. This conclusion is at variance with
paleobotanical determinations of previous investigators.

Waters and Krauskopf (1941, p. 1372) tentatively assigned the sedimentary rocks to the Oligocene on the
basis of identification of plant fossils sparsely distributed throughout the arkosic facies. In 1965, a new
collection of plant fossils was made from an arkosic
bed east of Shankers Bend by J. A. Wolfe of the U.S.
Geological Survey. His collections (USGS Paleobot.
locality 11014-7; NW¼ sec. 13, T. 40 N., R. 26 E.)
yielded several varieties useful for dating the rocks;
he reports as follows (written communication, 1968):

Fossils

Equisetum sp.

Metasequoia sp.

Osmunda sp.

Alnus n. sp., aff. A. cuprovalis Axel.

Cercidiphyllum crenatum (Ung.) R. W. Br.

Euptelea n. sp.

cf. Spondias sp.

The material from locality 11014-7
contains little that is diagnostic as to age.
Euptelea has not been recorded in beds
older than Eocene and the youngest known
record in western North America is in beds
of late Eocene age, although the genus is
still extant in Asia. The Cercidiphyllum
was at one time thought to indicate a middle
Oligocene or younger age (Wolfe, 1968),
but recent unpublished data indicate that
the Cercidiphyllum crenatum phylad was
present in Paleocene time. In addition,
"Cercidiphyllum" piperoides which was
thought to be the early Paleogene ancestor
of C. crenatum is now thought to represent
Tetracentron. The species of Alnus is related
to a late Eocene species that occurs in later
Eocene floras such as the Copper Basin and
Republic. Perhaps the most significant species
is the Metasequoia, which has the wide,
blunt needles characteristic of the Paleocene
and early Eocene Metasequoia occidentalis
(Newb.) Chan. If the known lower age limit
of Euptelea is given weight, then the overlap
of that range with the Metasequoia would
indicate a probable early Eocene age.

Wolfe's findings are compatible with the radiometric ages of the crosscutting dacite plugs and indicate
that the sedimentary rocks are probably early Eocene in
age.

Interpretation of Tertiary
Sedimentation

The source of the conglomerate and possibly the arkosic facies lies several miles to the northwest within the Similkameen plutonic complex. The great thickness of the conglomerate, the coarse nature of its cobbles, its easterly interfingering with arkose, and its position with respect to the area underlain by the Similkameen composite pluton suggest that the conglomerate is a lithified alluvial fan that was deposited at the foot of hills that rose to the northwest. The conglomerate was, at least in part, probably deposited subaerially, judging by the poor sizing and general lack of bedding. The arkosic facies, on the other hand, is better sized and shows good bedding and abundant ripple marks, suggesting predominantly subaqueous deposition, probably deltaic or bottom sediment within a lake occupying the Tertiary basin.

It is likely that deposition within the basin was accompanied by uplift of the source area, thus creating the gradients necessary to move cobbles the size of those found in the conglomerate. The abrupt upward transition from metamorphic to granitic detritus may be due to shifts in the drainage pattern or, less likely, may signal the deroofing of the Similkameen plutonic complex.

Deposition of the arkosic member ultimately ceased and was followed by a brief period of erosion. An abrupt onset of vulcanism and renewed sedimentation within the basin buried the older beds beneath a cover of pyroclastics and clastics containing volcanic detritus.

IGNEOUS ROCKS

Isolated masses of porphyritic hornblende dacite are found as small plugs and flows irregularly scattered along a belt parallel to the valley of the Okanogan River. Parts of two such masses, both plugs, lie within the Loomis quadrangle, and because they are more resistant than the Tertiary clastic rocks that they intrude they crop out as hills with sloping irregularly planar summits and steep scarplike sides.

Lithology

The rock is typically light pinkish gray with scattered acicular subhedral hornblende phenocrysts in a fine-grained to aphanitic matrix. The hornblende crystals are characteristically aggregated into irregular rosettes or clots. Quartz is neither megascopically nor microscopically visible, but an X-ray diffraction pattern showed it to be a fairly abundant constituent. A specimen of the typical rock, examined microscopically, is composed of approximately 40 percent equant plagioclase (andesine-labradorite) phenocrysts; 10 percent hornblende laths; 5 percent minute crystals of augite, yellowish-orange chlorite? (probably after biotite), apatite, and magnetite; and a weakly birefringent unresolvable matrix. The index of refraction of a fused sample indicates a probable silica percentage of about 63, using the average silica refractive-index curve of Huber and Rinehart (1966, fig. 7).

Age

Potassium-argon ages of hornblende from samples of three hornblende-bearing dacite masses have been determined. These include the plug half a mile north of the Similkameen River (sample L-590, $SE\frac{1}{4}$ sec. 12, T. 40 N., R. 26 E.), the plug a mile south of the Similkameen River (sample L-147, $SE\frac{1}{4}$ sec. 24, T. 40 N., R. 26 E.), and a flow capping Whitestone Mountain about 4 miles southeast of Enterprise in the Oroville quadrangle (sample L-657, $NW\frac{1}{4}$ sec. 21, T. 38 N., R. 27 E.). The determinations (see p. 63) were made by John Obradovich of the U.S. Geological Survey in Denver, Colo. (written communication, 1967).

The potassium-argon ages of the three masses are therefore early Eocene and apparently mutually indistinguishable within the limits of analytical error. Mathews (1964) found that the potassium-argon ages of similar rocks in southern British Columbia fall within the interval 45-53 m.y. The hornblende-dacite plugs and flows of the Loomis area appear to be a southward extension of this volcanic province, judging by the potassium-argon ages.

Analytical Data[*]

Sample	K (percent)	$Ar^{40} \times 10^{-11}$ (moles/g)	rad. Ar^{40} (percent)	Age (m.y.)
L-590	0.714	6.61	71.7	51.4 ± 2.6
L-147	0.752[**]	7.06	73.9	52.1 ± 2.3
L-657	0.673	5.96	83.1	49.1 ± 1.8

[*] Determinations by John Obradovich, U.S. Geological Survey.

[**] Single isotope dilution value; all others in duplicate.

METAMORPHISM

ANARCHIST GROUP AND KOBAU FORMATION

The general metamorphic grade in the Loomis quadrangle is greenschist facies with zones of amphibolite facies somewhat irregularly distributed along contacts with plutonic rocks. No extensive penetrative deformation has accompanied metamorphism because primary textures are fairly well preserved; in fact, bedding locally is enhanced rather than obscured. Schistosity is most pronounced in the bedded rocks and is typically parallel to bedding planes; it is weakly developed or absent in unbedded rocks.

Except near contacts with granitic intrusives, the following are typical assemblages of essential minerals:

Pelitic rocks—quartz, albite, K-feldspar, muscovite, biotite

Calcareous and siliceous rocks—recrystallized calcite and quartz

Mafic igneous rocks—quartz, albite, K-feldspar, biotite, muscovite, chlorite, actinolite, epidote/clinozoisite, zoisite

The mafic igneous rocks provide the best indications of metamorphic grade and clearly indicate greenschist facies. Although granoblastic texture and schistose structure are commonly present in some degree, relict textures abound, and the texture most common in the mafic igneous rocks is blastodiabasic. Relict clastic texture is generally recognizable in all but the finest grained metasedimentary rocks.

Although hornfelsic texture is typical near contacts with granitic rocks, mineral assemblages in metamorphosed mafic igneous and calcareous rocks clearly indicate that the highest grade attained is in the amphibolite facies of regional metamorphism. Diagnostic assemblages include the minerals anthophyllite, cummingtonite, and diopside; other minerals present that are compatible with, but not indicative of, this grade include andesine, hornblende, actinolite, biotite, muscovite, epidote/clinozoisite, zoisite, garnet, wollastonite, prehnite, and spinel. The common presence of epidote/clinozoisite with plagioclase more anorthitic than albite indicates that conditions of regional rather than contact metamorphism prevailed; that is, that pressures operative were above those generally considered typical of contact metamorphism. Widespread occurrence of the same mineral pair— epidote/clinozoisite and plagioclase—further indicates that the highest subfacies of the amphibolite facies generally was not reached.

In his treatise on metamorphic petrogenesis, Winkler (1965, p. 106) compares the Abukuma facies series of Miyashiro (1961) with the classical Barrovian facies series of Scotland. He concludes that the conditions which produced the Abukuma facies lie between those for the hornfels facies of contact metamorphism and the Barrovian facies of regional metamorphism. The difference is presumably a function of depth of burial during metamorphism—hornfels facies reflects shallow depth, Barrovian facies considerable depth. In the metamorphic terrane of the Loomis quadrangle many mineral assemblages suggest affinity with the Abukuma facies as shown by the following examples: (1) Former presence of andalusite in the Bullfrog Mountain Formation at both Douglas and Aeneas Mountains, indicated by euhedral square prisms pseudomorphed by sericite, is compatible with hornfels and Abukuma facies, not Barrovian facies. Occurrence of the pair plagioclase-epidote, in similar environment but in different rocks, rules out hornfels facies. (2) At Palmer Mountain an increase in grade to the west, from greenschist facies on the east, is shown by an increase in anorthite content of plagioclase from albite to andesine, and by the presence of green pleochroic hornblende instead of actinolite as the only amphibole. Chlorite and epidote/clinozoisite occur as apparently stable minerals in both assemblages. The plagioclase-epidote-hornblende-chlorite assemblage is stable only in the highest part of the greenschist facies of the Abukuma type—it is not a stable assemblage in either Barrovian or hornfels facies. (3) At Lone Pine Creek a metamorphosed greenstone of the Kobau Formation near a contact with shonkinite shows the following assemblage: andesine-hypersthene-anthophyllite-cummingtonite-biotite-spinel. The occurrence of orthopyroxene with amphibole is diagnostic of the highest subfacies of the Abukuma amphibolite facies, designated the orthopyroxene-hornblende subfacies (Winkler, 1965, p. 106).

Strong indications that conditions overlapped those ascribed to both Barrovian and Abukuma environments are provided by mineral assemblages in the Kobau Formation and the Palmer Mountain Greenstone near granitic contacts. Both contain the assemblage—quartz, andesine, hornblende, diopside, epidote/clinozoisite—typical of the low or medium part of the Barrovian almandine-amphibolite facies, and not a stable assemblage of either the hornfels or Abukuma facies. Chlorite occurs locally with the above assemblage but petrographic relations suggest that it is a later alteration product.

Wollastonite, long considered a mineral typical of contact metamorphism, is fairly common near granitic contacts in the calcareous rocks of the quadrangle. Experimental work by Harker and Tuttle (1955a, b) and Greenwood (1962), however, indicates that pressures and temperatures at which wollastonite will form are highly dependent on the partial pressure of CO_2; hence without some knowledge of the latter, wollastonite is of little value as a facies indicator.

ELLEMEHAM FORMATION

Primary textures in the Ellemeham Formation are only slightly affected by metamorphism. In fact, delicately recrystallized spherulites and textures resembling devitrified glass are commonly present in the greenstones of the basal Ellemeham, and are only locally obscured by hornfelsic growth of fine-grained metamorphic minerals.

The common mineral assemblage in the greenstone of the Ellemeham is quartz-albite-muscovite-chlorite-epidote/clinozoisite, equivalent to the lowest subfacies of the Barrovian greenschist or to the albite-epidote-hornfels facies. Orange-brown biotite occurs sporadically and presumably reflects slight variations—probably temperature—in the conditions of metamorphism. The rocks within a few hundred yards of the alkalic rocks associated with the Shankers Bend diatreme are an exception in that the assemblage quartz-plagioclase-K-feldspar-hornblende-diopside-biotite-epidote/clinozoisite-garnet is common, indicating an increase in grade to the intermediate subfacies of the Barrovian almandine-amphibolite facies.

We conclude that throughout most of their extent the rocks of the Ellemeham Formation are significantly less metamorphosed than the subjacent rocks of the Kobau. This circumstance appears to require metamorphism of the Kobau Formation, Anarchist Group, and Palmer Mountain Greenstone prior to deposition of the Ellemeham. Thus we postulate two metamorphic events whose ages can be bracketed only within broad limits. Rocks of the Anarchist Group, Kobau Formation, and Palmer Mountain Greenstone were metamorphosed during the earlier event, and the rocks of the Kobau Formation were again metamorphosed along with those of the Ellemeham Formation during the later event. A maximum age for the earlier event is fixed by fossils of Late(?) Permian age found in the Anarchist Group. Because the younger but undated Kobau Formation, lying unconformably above the Anarchist, was also involved in this event, it is reasonable to assume that metamorphism may have occurred as late as Early or Middle Triassic. This early metamorphism preceded deposition of the Ellemeham Formation.

The later metamorphism may be related to the intrusion of the Similkameen composite pluton because alkalic rocks, which are lithologically similar to those associated with the Similkameen, have intruded and metamorphosed the Ellemeham near Shankers Bend. On the other hand, the Similkameen pluton contains features that can be interpreted as evidence that it, too, has been mildly metamorphosed. Such features include the widespread occurrence of epidote generally in amounts of several percent, and the common occurrence of incipient to mild cataclastic texture. So, although the time relations between the later metamorphism and the Similkameen are ambiguous, the amount of time between the two episodes of metamorphism must have been considerable—sufficient at least, to permit erosion of the Anarchist and Kobau terrane, prior to deposition of the Ellemeham, to the

considerable depth required to expose rocks metamorphosed at pressures of the Barrovian and Abukuma facies series.

STRUCTURAL GEOLOGY

The structural features of the rocks in the Loomis quadrangle record a history of deformation extending from the end of the Permian to the early Tertiary. The folds, which are found exclusively in the metamorphic rocks, range from minor crenulations to folds with wave lengths of several miles. The axial traces of the major folds (fig. 26) describe, in general, broad arcs that are convex to the northeast and that strike roughly northwest. The minor fold axes do not parallel the axes of the northwest-trending major folds.

Several faults offset the rather thick stratigraphic units shown on the geologic map. With few exceptions, the high-angle faults trend northwest to northeast. The

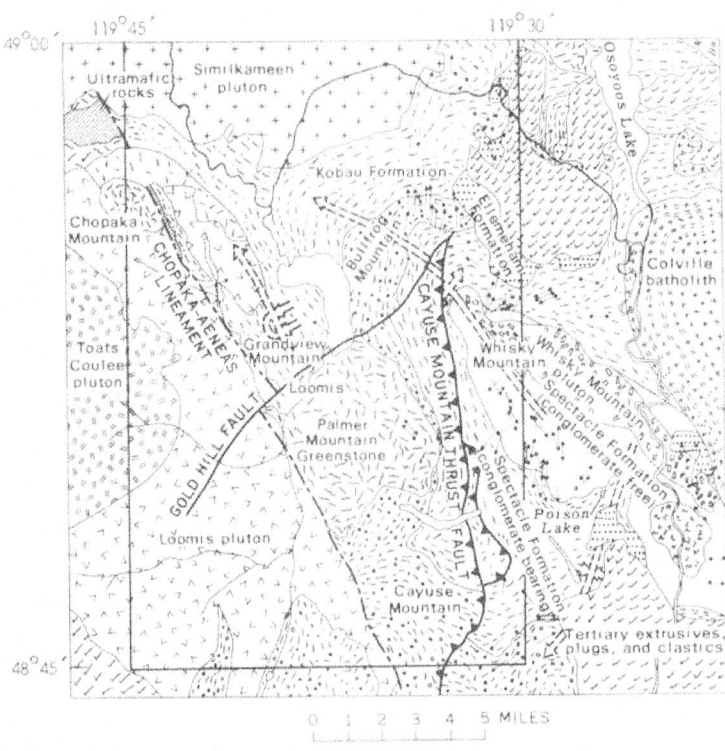

FIGURE 26.—Map showing main structural elements within the Loomis quadrangle and contiguous areas. Axes of major folds, major faults, trend of horizontal minor folds (double-pointed arrow) and trend and direction of plunge of inclined minor folds (arrow) are shown.

single reliably known low-angle fault of consequence trends about north, and probably is a thrust, with the west block overriding the east block.

MAJOR FOLDS

Wannacut Lake Area

The Whisky Mountain anticline forms the most conspicuous structural feature of the east-central portion of the Loomis quadrangle. The fold is a large, rather symmetrical northwestward-plunging anticline which extends southeastward into the adjacent Oroville quadrangle. The southwest limb dips to the southwest at angles of 30°-75°, locally shallowing to 10°, and can be easily traced northward from Spectacle Lake by following the prominent limestone beds of the Spectacle Formation to Bullfrog Mountain. In the northeast limb, dips are as steep as 90° at the quadrangle boundary. Dips are gentle near the axis, which plunges northwest 25°-35° in the Hicks Canyon area, where the most reliable measurements were made. The plunge seems to steepen as the axis curves more westwardly through Ellemeham Mountain. A prominent massive metachert member of the Kobau Formation along the northwest extension of the axial trace forms a dip slope of 35°-40° on the crestal portions of Ellemeham Mountain, but steepens and becomes nearly vertical to the northwest. Beds farther to the northwest, near the contact with the Similkameen composite pluton, are probably overturned, since they dip steeply to the southeast. The apparent overturning of the northwest extension of the anticline may be due to deflection of the beds by forcible injection of the Similkameen pluton.

The core of the anticline is pierced by the Whisky Mountain pluton. A few attitudes taken near the contact southeast of Wannacut Lake dip to the northeast, suggesting that the anticlinal axis actually lies to the southwest of the granodiorite. If so, the long axis of the anticline diverges slightly from the long axis of the pluton.

The age of the fold can be established within broad limits, since it is cut by both the Similkameen and the Whisky Mountain plutons and involves strata as young as the Kobau Formation. The Kobau, forming the northeastern flank of the fold, is overlain by unfolded Ellemeham along a pronounced angular unconformity, indicating that the fold substantially predates deposition of the Ellemeham.

Chopaka and Grandview Mountain Areas

The metamorphosed rocks of the Kobau and Palmer Mountain Formations and the Anarchist Group, located west of Palmer Lake and the Similkameen River, are cut off and embayed by younger plutonic rocks. The attitude of the strata within these embayed remnants is grossly concordant, in spite of certain irregularities, and the rocks trend northwest and dip steeply to the southwest. On the eastern slope of Grandview Mountain, the Kobau dips homoclinally to the west-southwest at angles of 45°-60°. Near the crest of Grandview Mountain the strike of the beds is quite varied, and dips are generally steep. A few massive chert beds can be followed along strike in this area, but only for short (1/4 to 1/2 mile) distances. Their undulations, combined with the diverse attitudes of enclosing strata, indicate considerable deformation. Farther west, along the east side of Chopaka Creek the attitudes are generally steep, dipping chiefly to the west, and locally to the east.

Intensely deformed chert conglomerate of the Anarchist Group was found lying west of the Kobau Formation at two localities. The Anarchist is in contact with the Kobau near Chopaka Creek (sec. 3, T. 39 N., R. 25 E.). The presence of the Anarchist at this locality is difficult to reconcile with its overall distribution and stratigraphic position below the Kobau without postulating either a fault of considerable displacement, or, more plausibly, a synclinal axis to the east. Relict conglomeratic texture was recognized in high-grade, partly migmatized metamorphic rock lying 1½ miles west of the quadrangle border (sec. 14, T. 38 N., R. 24 E.), indicating that this rock, too, is Anarchist.

These admittedly sketchy bits of evidence suggest that the Kobau remnants are flanked by remnants of older metamorphics to the west. If this conjecture is correct, then the overall configuration of the Kobau is probably that of a tightly appressed, overturned syncline whose axis trends northwest, paralleling the axis of the Whisky Mountain anticline. Regardless of the precise nature of the structure suggested by the distribution of Anarchist and Kobau remnants, this structure is clearly cut off by the Loomis pluton; hence it formed during or prior to Late Triassic (the age of the Loomis pluton), but not before Late(?) Permian (the age of the Anarchist Group).

Cayuse Mountain Area
(T. 38 N., R. 26 E.)

The homoclinal sequence forming the north-striking west limb of the Whisky Mountain anticline in the Rainbow Lake area (secs. 22, 23, T. 39 N., R. 26 E.) was traced southward across the Loomis-Spectacle Lake Valley and into Cayuse Mountain. Within the Cayuse Mountain area, the strike swings abruptly to the east and the beds

are overturned with moderate to steep northerly dips. There, the conspicuous east-trending limestone beds of the Spectacle Formation give sharp definition to the overturned limb.

Bedding tops found in the Bullfrog Mountain Formation west of the thrust fault and south of the limestones (secs. 22, 23, 26, T. 38 N., R. 26 E.) generally confirm the overturned character of the strata, although reversals are present. East of the thrust fault (in sec. 24) most of the bedding tops inferred from sedimentary features and bedding-cleavage relations corroborate overturning.

The pattern of major folding is considerably obscured by later thrust faulting, and the interpretation of the folding depends on the somewhat tenuously inferred configuration of the faults. If the undulating, subhorizontal limestones (in secs. 13, 14, T. 38 N., R. 26 E., and sec. 35, T. 39 N., R. 26 E.), instead of forming the middle plate of an imbricate thrust zone as hypothesized, are part of the stratigraphic sequence of

FIGURE 27.— Small-scale chevron folds in metagraywacke of the Spectacle Formation; pocket-knife for scale.

the lower plate, then a more complex fold is required to accommodate them. A plausible alternative would be to correlate the flat-lying limestones with the overturned rib to the south (secs. 25, 26, T. 38 N., R. 26 E.) and postulate an overturned nappelike anticline with a northeast-trending axis located between them (in sec. 24).

Reconnaissance in the Poison Lake area of the Oroville quadrangle to the east revealed a prominent syncline trending north-northeast and plunging moderately to the south-southwest (fig. 26). The overturned beds of the Cayuse Mountain area strike toward the axis of the syncline and probably swing northeast to form its western limb, although overlap by Tertiary beds precludes confirmation of this hypothesis. The syncline appears to have been impressed on and is thus younger than the Whisky Mountain anticline, and is itself unconformably overlain by rocks provisionally correlated with the Ellemeham Formation. These relations suggest that the east- to northeast-trending overturned beds of the Cayuse Mountain area are a limb of a north-northeast-trending superimposed crossfold, a result of deformation occurring after the deformation that produced the northwest-trending folds but prior to deposition of the Ellemeham.

FIGURE 28.—Asymmetrical minor folds in limestone of the Spectacle Formation; hammer for scale.

MINOR FOLDS AND CRENULATIONS

The bedded rocks of the Anarchist Group, the Palmer Mountain and Kobau Formations, and, rarely, the Ellemeham Formation locally show small-scale folds and crenulations (fig. 27). The folds commonly have amplitudes of a few inches and wavelengths of a foot or so, but range in size from microcrenulations that are detectable only with the aid of a hand lens to flexures with wavelengths of several tens of feet (fig. 28). In places the folds are obviously parasitic on larger folds. At one locality (sec. 6, T. 38 N., R. 27 E.), folds of three orders of magnitude can be seen: the largest with a wavelength of about 75 feet and an amplitude of 25 feet; superimposed on this is an asymmetrical crenulation with a wavelength of 2 to 24 inches and amplitude of 1 to 6 inches; this crenulation, in turn, shows a microcrenulation of varied size, but commonly with a wavelength of about one-tenth of an inch and an amplitude of about one-thirtieth of an inch.

A stereographic plot of the axes of minor folds and crenulations (fig. 29) shows a concentration of horizontal to moderately plunging axes trending north to northeast and other lesser but well-defined concentrations. The map (fig. 26) indicates considerable regularity in attitude of fold axes in local areas with gradual to abrupt transitions of the dominant attitude from one area to another. Refolded crenulations or minor folds were not detected in the field, even in the areas where the dominant fold attitude changes abruptly.

Where symmetrical, the crenulations typically are sinusoidal in cross section, and they approach it where asymmetrical, but sharply plicated chevron folding was also observed. The chevron folds grade to kink folds showing a well-developed axial plane cleavage not visibly marked by parallel recrystallization of minerals. The cleavage resembles a close-spaced fracture cleavage, and in certain areas (as in the eastern part of Palmer Mountain) can be detected outside the area in which crenulations are abundant.

The minor folds and crenulations are concentrated within a belt along the eastern edge of the quadrangle (see fig. 26) and are mostly confined to rocks older than the Ellemeham Formation, as for example, north of Whisky Mountain. There, the abundant crenulations in the Anarchist have no counterpart in the unconformably overlapping Ellemeham. Several exceptions to this generalization must be noted. A few minor folds were mapped within the Ellemeham in the Shankers Bend and the diatreme areas (secs. 12, 13, 14, T. 40 N., R. 26 E.) and within correlative rocks adjacent to the Colville batholith in the Oroville quadrangle (fig. 26). Even in these areas, minor folds in the Ellemeham are rare and are only very weakly developed.

The foregoing evidence indicates that the majority of the minor folds and crenulations resulted from one or more deformations that occurred prior to the deposition of the Ellemeham Formation. The problem remains as to whether the minor folds were produced in a single period of deformation or in several, and whether the deformation(s) are related to that which produced the major northwest-trending folds (Whisky Mountain anticline) or the major north-northeast-trending crossfold (in the Poison Lake area). The absence of refolded minor folds and the gradual transition in attitude, particularly at the junction of areas showing markedly different trends, suggests that a single main episode of regional deformation is responsible for the minor folding. Assuming that the magnitude and extent of post-Ellemeham folding is negligible, the various concentrations of fold axes (fig.

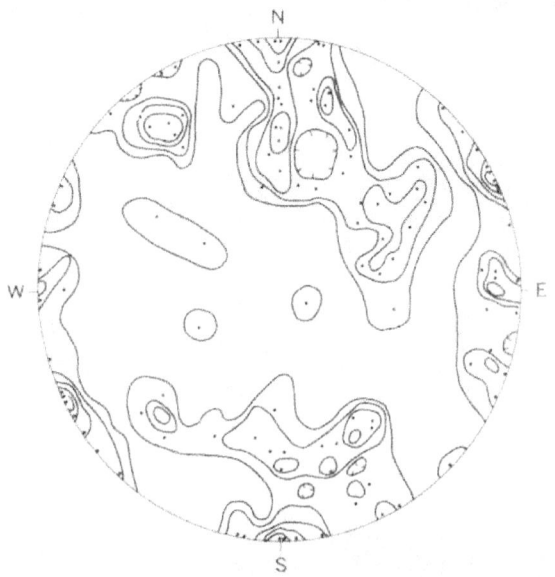

FIGURE 29.—Lower hemisphere equal-area stereographic projection of axes of 91 minor folds and crenulations in the Loomis quadrangle and contiguous areas. Contours at 1, 2, 3, 4, and 5 percent; maximum 6 percent.

29) therefore probably reflect the differing local responses of an anisotropic medium during a single widespread period of deformation.

The minor folds clearly lie across the axis of the Whisky Mountain anticline and, as noted, lie within a north-trending belt, suggesting that they are superimposed on the northwest-trending major folds of inferred Triassic age. On the other hand, none of the folds on the north-northeast-trending syncline in the Poison Lake area appear to be crossfolds. In fact, those located within the overturned strata in the Cayuse Mountain area and those northwest of Poison Lake (fig. 26) appear to be parasitic on the syncline. These circumstances suggest to us that the minor folds and crenulations were caused by the same deformation that created the syncline.

PRE-TERTIARY FAULTS AND LINEAMENTS

Chopaka-Aeneas Lineament

The Chopaka-Aeneas lineament is defined by the roughly aligned trace of the contact between granitic and metamorphic rocks on Chopaka, Grandview, and Aeneas Mountains (pl. 1 and fig. 26). Additional elements of the lineament on Chopaka Mountain are the elongate serpentinite body; farther north, the contact between the Palmer Mountain Greenstone and the Kobau Formation; still farther north, in the Horseshoe Basin quadrangle (fig. 26), the contact between the Kobau and ultramafic rocks. The lineament trends north-northwest, extends nearly the full span of the quadrangle, and terminates against the Colville batholith in the Conconully quadrangle to the south (pl. 1, inset) and against the Similkameen composite pluton to the north.

Along most of its length, the contact of the Loomis pluton with the metamorphic rocks along the lineament is characterized by breccias that appear to be intrusive; the contact is ragged in detail, thereby supporting the interpretation of the breccia as intrusive rather than tectonic. On Aeneas Mountain, however, slickensides, fault breccias, and zones of mylonite along the lineament both in the Loomis pluton and in the Bullfrog Mountain Formation indicate strike-slip faulting of post-Loomis age and demonstrate that the lineament remained an active structural element subsequent to the intrusion and solidification of the Loomis.

The original nature of the lineament is obscured by the Loomis pluton, but by analogy to the more recent displacement on Aeneas Mountain it seems likely that the lineament was originally a strike-slip fault. If it is, maximum displacement probably is measurable in thousands of feet rather than miles, since the metamorphic rocks on either side are roughly correlative.

The lineament knifes through the folded metamorphics without substantial deviation, indicating that it is probably younger than the folding. Since its trace is occupied by an "alpine-type" ultramafic intrusive on Chopaka Mountain that is pre-Loomis in age, the lineament must be older than the Late Triassic Loomis. Thus the evidence suggests that the lineament developed prior to Late Triassic; it may have served as a gross control for the eastward spread of the Loomis magma, but after pre-Late Triassic folding.

Gold Hill Fault

The Gold Hill fault strikes northeast and dips to the northwest. The dip becomes progressively steeper from northeast to southwest, as shown by measurements of 40° on Bullfrog Mountain, 70° near Chopaka Creek, and 80°-85° on Gold Hill. On the west slope of Bullfrog Mountain the fault is marked by a zone of fault breccia and vein quartz, and its exposure in Rattlesnake Draw is conspicuously marked by a wide zone of mylonite and breccia. Farther southwest, between Rattlesnake Draw and Toats Coulee, the fault cannot be identified in outcrop, although pronounced jointing coincident with its southwestward projection is apparently responsible for the incision of a deep gully on the south slope of Rattlesnake Mountain. On Gold Hill the fault apparently splits into several parallel en echelon high-angle faults carrying brecciated quartz and sulfides in the fault zones. Slickensides indicate strike-slip movement.

The fault appears to offset the Aeneas-Chopaka lineament in a right-lateral sense, though critical exposures at the zone of intersection are lacking. The contact between the Bullfrog Mountain Formation and the Spectacle Formation (east of Palmer Lake) also appears to be offset in a right-lateral sense, although here, too, the presumed offset cannot be verified because of poor exposure.

The presumed trace of the two-mile segment at the northeast end of the fault, is based on a weak lineament which was observed on aerial photos but cannot be traced northeast of its projected intersection with the thrust fault in Hicks Canyon (sec. 34, T. 40 N., R. 26 E.).

The fault zone, where exposed in a roadcut on the Gold Hill road (sec. 4, T. 38 N., R. 26 E.), is occupied by diabase dikes that are themselves slickensided. Their presence, together with the previously mentioned presence of brecciated quartz veins at Gold Hill and Bullfrog Mountain, indicates a history of recurrent movement along the fault beginning after solidification of the Loomis magma and possibly extending into Cenozoic time.

Cayuse Mountain Thrust Fault

The Cayuse Mountain thrust fault trends north-south (pl. 2 and fig. 26) and dips shallowly to the west. Three-point solutions of the dip of the fault indicate that it dips 11°-13° in the Cayuse Mountain area and steepens to 25° west of Wannacut Lake along the northern segment of the fault.

The fault line has good topographic expression in the uplands of the Cayuse Mountain area and the Rainbow Lake-Wannacut Lake area because it is marked by a continuous line of shallow, alluvium-filled gullies and linear depressions. Rocks adjacent to the fault trace are commonly sheared and brecciated or much fractured and highly jointed.

The actual fault can seldom be observed, owing to alluvial cover, but at one locality northwest of Wannacut Lake, (sec. 3, T. 39 N., R. 26 E.) the trace of the fault is marked by a zone about 100 feet thick of brecciated to mylonitic Anarchist conglomerate and other finer grained clastics, and rocks of the adjacent Whisky Mountain pluton are variously sheared and mylonitized and may be cut off by the fault.

The main thrust fault is flanked to the east by subsidiary thrusts with shallower westward dips. The most important, and unfortunately the least well documented, of the thrusts crosses the area both north and south of Spectacle Lake. The upper, overthrust, plate contains undulating but grossly horizontal Anarchist limestone and minor interbedded metaclastics that apparently are thrust over steeply dipping metaclastics of the Spectacle Formation. The thrust plane dips 12°-15° W. where it passes beneath the limestone that caps hill 3093, 1½ miles north of Spectacle Lake. There the limestone appears to be tightly folded into a recumbent anticline whose axis strikes northeastward; it contains a lens of phyllite sandwiched between the two limbs along the shallowly westward-dipping axial plane. A layer of similar limestone south of Enterprise (secs. 13, 14, T. 38 N., R. 26 E.) also is presumed to be part of the overthrust plate since its much folded but subhorizontal attitude is difficult to reconcile with the steeper dips of the rocks below.

The belt of overturned Anarchist rocks east of Horse Springs Coulee in the Cayuse Mountain area is offset and apparently thrust to the east, possibly as much as several miles along the main thrust. The offset along the northern part of the fault is apparently less, declining as the fault plane steepens, perhaps indicating a hinge at the northern end.

Several other low-angle faults of less importance were found within the quadrangle. One that is worthy of mention offsets rocks of the Ellemeham Formation against those of the Anarchist, a mile northeast of Hicks Canyon (secs. 26, 27, T. 40 N., R. 26 E.). The fault strikes east and dips 25°-40° to the north. It is marked by a 5- to 25-foot-wide zone of brecciated and strongly jointed rock adjacent to the fault.

A large thrust fault located 12 miles south of the Loomis quadrangle was previously described by Misch (1951), who concluded that the fault was younger than the major northwest-trending folds but was itself deformed by subsequent folding. The Cayuse Mountain thrust fault is probably younger than both the major northwest-trending folds and the later north- to northeast-trending folds and crenulations. It also must postdate the Whisky Mountain pluton judging by the cataclasis of the granodiorite near the fault. If related to the low-angle fault at the base of the Ellemeham mentioned above, as is likely, then it is also younger than the Ellemeham.

TERTIARY WARPING AND FAULTING

Lower(?) Eocene clastic deposits along the north-east edge of the Loomis quadrangle and in adjacent areas of the Oroville quadrangle occupy an irregular north-trending structural basin. Beds in the southern half of the basin (south of the Similkameen River) dip centripetally toward a central northwest-trending axis, which lies just east of the Loomis quadrangle, but the beds in the northern half dip homoclinally to the east. Dips within the Loomis quadrangle along the western edge of the basin generally range from 20°–30° and are inclined to the east; dips up to 55° were observed in a small fault block just east of Shankers Bend and alongside the fault in Ellemeham Draw.

The Eocene sedimentary rocks and the dacite plugs which cut them are offset along steeply dipping north- and northeast-trending normal faults. The faults probably have maximum displacements of several hundred feet, but can be traced for only short distances on strike into the metamorphic rocks. This suggests that the faults were localized in and near the basin, and therefore the faulting and downwarping are related.

The Tertiary deformational cycle probably began with downwarping of the basin, followed by or perhaps concomitant with sedimentation, then vulcanism, and lastly, faulting. These phases undoubtedly overlapped, and rather than marking discrete episodes, they serve to identify only the main events during a continuous period of deformation.

INTERPRETATION OF THE
STRUCTURAL RECORD

The structures described in the previous paragraphs suggest repeated episodes of deformation beginning in the Triassic and continuing into the Tertiary. The northwest-trending folds and faults and subsequent or possibly concomitant metamorphism and plutonism attributed to the Triassic account in general for the basic configuration of the rocks of the Loomis area. General similarity in fold trends and correspondence in age leave little doubt that these features are the local expression of widespread orogeny (White, 1959, p. 72). The

regional maximum principal stress axis was evidently oriented northeast-southwest, presuming that it can be equated with the axis of maximum compression of Loomis rocks, that is, perpendicular to the axes of folds.

Tectonism subsequent to the Triassic orogeny produced the minor folds and the overturned crossfold of the Cayuse Mountains, but was probably less intense than in the Triassic. Clustering of minor-fold axes around a north-northeast direction suggests that the regional maximum principal stress axis for these structures was oriented west-northwest. This episode of deformation is provisionally assigned to the Jurassic and Cretaceous because of the general parallelism of these folds with those attributed by Yates and others (1966) to the Jurassic and Cretaceous in northeastern Washington.

Thrusting along the north-south-trending Cayuse Mountain fault apparently requires a further rotation of the regional axis of maximum principal stress to a west-east orientation. Evidence from the Loomis quadrangle indicates only that the thrusting was later than the north-northeast-trending family of folds, perhaps substantially later, if, as seems likely, the thrusting is younger than the Ellemeham. Lacking more definitive field relations, we tentatively correlate the thrusting with the extensive thrusting of Cretaceous age to the west recognized by Misch (1966).

A provisional sequential arrangement of the structures of the Loomis quadrangle along with the probable orientation of the regional axis of maximum principal stress is listed in table 6. It would appear that the axis of maximum principal stress shifted from a southwest orientation in the Triassic, to west-northwest during the Jurassic and Early Cretaceous, and then to west in the Late Cretaceous and Tertiary.

PHYSIOGRAPHY AND GLACIATION

The terrain of the Loomis quadrangle is mountainous, with irregular hills along the east side rising westward to rugged, deeply dissected mountains, which eventually merge with the high peaks of the Cascades some distance west of the quadrangle. In the central part of the quadrangle, summit areas reach elevations of 4,500 feet and are generally broad, gent-

TABLE 6. — Sequence of structural events in the Loomis quadrangle

Probable age of postulated event	Description of event	Direction of maximum principal stress
Triassic (pre-Loomis pluton)	Folding on northwest axes of Whisky Mountain anticline and similar structures. Establishment of Aeneas-Chopaka lineament.	Northeast-southwest.
Late Triassic and Early Jurassic	Plutonism (Loomis, Toats Coulee).	
Jurassic or Cretaceous (pre-Ellemeham Formation)	Folding on north-northeast axes (minor folds and the cross-fold of the Cayuse Mountain area). Strike-slip movement on Aeneas-Chopaka lineament. Emplacement of Colville batholith (east of the Loomis quadrangle).	West-northwest - east-southeast.
Epeirogenic uplift, erosion, extrusion of lavas of Ellemeham Formation		
Jurassic and Cretaceous (post-Ellemeham)	Plutonism (Similkameen).	
Jurassic and Cretaceous (post-Similkameen)	Strike-slip movement on northeast-trending faults (Gold Hill, Nighthawk).	East-west(?).
Cretaceous	Overthrusting on north-south-trending thrusts (Cayuse Mountain thrust fault).	East-west.
Tertiary	Warping, normal faulting.	East-west(?).

ly convex, and without well-defined peaks. To the west, slopes are more precipitous and the summit areas are narrow ridges and peaks.

Drainage in the area is dominated by the Similkameen River and its now-abandoned ancestral channel. The abandoned channel and the northern part of the channel now occupied by the river (fig. 30) constitute a broad, steep-walled trench that runs north-south the length of the quadrangle and whose U-shape cross-profile indicates previous occupancy by a valley glacier. The southern two-thirds of the trench is presently occupied by Sinlahekin Creek, a grossly underfit, north-flowing stream that empties into Palmer Lake that is situated near the junction of the Sinlahekin and the Similkameen. There the Similkameen leaves its broad glaciated canyon, turns abruptly from its general south-

FIGURE 30.—U-shaped valley of the Similkameen River, looking northward from the east flank of Chopaka Mountain.

ward course, and plunges northeastward through a rela-tively narrow canyon toward its confluence with the Okanogan River at Oroville, 10 miles to the east. This curious feature was observed by Willis (1887), who correctly deduced that the trench now occupied by Sinlahekin Creek was the ancestral course of the Similk-ameen River. After being occupied by a vast glacier, the trench was successively dammed by either drift or remnants of the glacier itself, causing diversion of melt water through, and consequent erosion of, the canyon now occupied by the eastward-flowing segment of the Similkameen River and also the east-west canyon now occupied by Spectacle Lake (fig. 31). Termination of the glacial epoch left the Similkameen River perma-nently flowing in its northern melt-water cutoff.

The very low gradient between the outlet of Palmer Lake and the Similkameen River is strikingly emphasized by a delta built into the lake, from its outlet. During heavy runoff, the level of the Similka-meen occasionally exceeds that of Palmer Lake, revers-ing the flow in the connecting channel and causing the outlet to become a temporary site of deposition.

The general extent and nature of glaciation in the Okanogan region is now fairly well known through the studies of Willis (1887), Dawson (1898), Smith and Calkins (1904), Daly (1912), Flint (1935), Waters

(1937, 1939), Nasmith (1962), and other workers. The Loomis area, in common with much of the Okanogan region, was overridden by a thick southward-flowing mass of ice, termed the "Okanogan lobe" (Flint, 1935, p. 173), which pushed somewhat farther south than the present location of the Columbia River. The general period of glaciation has been correlated with the Wis-consin Glaciation of the mid-continent, although a satisfactory detailed chronology has yet to be determined (Fryxell, 1960, p. 2060).

Daly (1912, pt. 2, p. 592) found striations and glacially polished bedrock as high as 7,200 feet on Mount Chopaka. Most of the lower summit areas are in part mantled with drift, and bedrock exposures are commonly striated and polished. Thus, the ice sheet in the Loomis quadrangle at its maximum must have been at least a mile thick over the present valleys. Striations are generally oriented between south and S. 30° E., confirming the reported regional southerly flow.

Subsequent melting and retreat of the ice sheet resulted in the release of large volumes of melt water and liberation of the load of detritus with which gla-ciers are charged. The effect on the topography by detrital accumulations and melt-water runoff was prob-ably as profound as earlier erosion by the ice itself. In addition to reorienting the master streams of the area,

FIGURE 31.—View of Spectacle Lake, looking westward from southwest of gauging station. Lake occupies presumed melt water diversion channel cut by ancestral Similkameen River during the Pleistocene Epoch.

as in the case of the Similkameen River, the melt-water streams cut into bedrock adjacent to the downwasting ice masses, eroding discontinuous, narrow, sometimes stairstepped channels and notches. Much of the detritus, consisting mostly of poorly stratified gravelly silt, was deposited as kames and terrace deposits and irregular discontinuous veneers. The terraces are usually gently sloping local surfaces, without correlative surfaces in contiguous areas, suggesting that the terraces formed in ponded areas alongside wasting ice rather than as extensive stream or lake deposits.

Some of the most conspicuous kames are perched near the summits or at high elevations along the flanks of the mountains (SE¼ sec. 21, T. 39 N., R. 26 E.; NW¼ sec. 9, T. 39 N., R. 26 E.). Deposits such as these were probably formed as the general melting of the ice sheet bared crestal areas of the mountains and ridges. This interpretation agrees with that of Nasmith

(1962), who hypothesized that the ice sheet melted by general downwasting of its upper surface rather than by retreat along a broad front.

The upper surface of some of the terracelike deposits, such as those mapped along the east-west leg of the valley of the Similkameen River northeast of Nighthawk, slope 10°-20° toward the axis of the valley. Where bedding is observed, it generally is grossly parallel to the upper surface. These features suggest that the deposits are fans rather than terraces and probably are composed of drift washed down from the higher slopes, perhaps both before and after recession of the glacier.

CHOPAKA MOUNTAIN LANDSLIDE THREAT

The summit area of Chopaka Mountain is a seriate ridge that towers approximately 6,000 feet over the Similkameen Valley, barely 2 miles to the east. The northeast face of that portion of Chopaka Mountain within the Loomis quadrangle (sec. 30, T. 40 N., R. 25 E.) slopes approximately 40° to the north-northeast. The other slopes fall more gently to the highlands to the south and west. A small part of the crestal area at the quadrangle boundary (in section 30) appears to be unstable. Here portions of the crest appear to be "calving off" along an east-west joint set. Large blocks of rock apparently are very slowly creeping downslope to the northeast, causing widening of the joints to veritable crevasses. The largest such fissure is about 10 feet wide, 75 feet long, and filled with rubble to within 10 to 20 feet of the surface. Adjoining blocks are successively offset downhill to the north (fig. 32).

There is no evidence of previous catastrophic movement, but considering the large volume of unstable

FIGURE 32.—Looking west at sparsely forested crest of Chopaka Mountain (left foreground). Tree-clad shoulder at lower right is presumed to have once been continuous with gently sloping crestal surface, but has moved downslope since the Pleistocene Epoch— through creep—to its present position.

rock now poised at the brink of the crestal area, the potential, at least for a landslide, is present. Fortunately, the area in the presumed path of such a landslide (sec. 20, SE$\frac{1}{4}$ sec. 16, and NW$\frac{1}{4}$ sec. 21, T. 40 N., R. 26 E.) is not densely populated. We recommend that the possible hazard be considered in future development.

MINERAL DEPOSITS

METALLIC DEPOSITS

The igneous and metamorphic rocks of the quadrangle are host to numerous small mineral deposits. Most deposits were at least superficially explored and a few were mined in earlier days, as attested by numerous shallow adits, shafts, and decaying headframes and cabins. None of the metal mines were producing in 1967, although development continues sporadically in a few of them. Most excavations have caved and are now inaccessible; thus the geologic circumstances of the deposits must be deduced from surface exposures and examination of the dumps, supplemented by descriptions in the geologic literature.

Vein Deposits

The majority of the prospects and mines within the quadrangle are located on quartz veins, most of which are of the fissure-fill type. A few small saddle-reef quartz veins were found in folded metamorphic rocks south of Wannacut Lake. Shear or fault zones that contain brecciated quartz and disseminated ore minerals are included in this category, as are veinlike massive sulfide replacement deposits (Copper World, Copper World Extension). Information pertaining to specific deposits is summarized in table 7.

The deposits are crudely zoned north to south (see pl. 2); the chief values are silver-lead in and adjacent to the Similkameen composite pluton, gold-silver-lead in the central part of the quadrangle, and gold-silver in the southern part of the quadrangle. The distribution

of massive sulfide deposits, such as those at Palmer Mountain, appears to be unrelated to the zoning.

In addition to gold, silver, and lead, accessory tungsten was found in a vein at the Four Metals mine in the western part of the northern zone, and accessory molybdenum occurs in some of the deposits west of Wannacut Lake.

Faulting has been important in the localization of the deposits, as many of the deposits are in shear zones. Faulting was probably concurrent with mineralization, for the ore minerals are commonly slickensided, brecciated, and healed.

Few generalizations can be made about the attitude of the veins. Considering the quadrangle as a whole, the veins display a bewildering array of strikes and dips (pl. 2), although within limited areas there is some consistency. The thickest quartz veins are the blanket deposits (Submarine mine) and the shallow-dipping (Triune mine) types, which reach thicknesses greater than 10 feet. Most of the veins are much thinner, and few have outcrop lengths greater than several hundreds of feet. These features and the failure of the Palmer Mountain and Ivanhoe adits to encounter, at depth, projected extensions of veins cropping out above suggest that the veins may not extend to great depth. Alternatively, the veins may have undetected rakes. Deposits associated with important structures (faults and shear zones) are likely to have greater persistence at depth.

Wall rock is altered adjacent to all the veins. Economic mineralization, however, seems to be restricted to the vein itself, at least in the area south of Wannacut Lake, where numerous analyses of both vein and wall rock were made. Wall rock alteration at most deposits is not megascopically apparent at distances greater than a few feet or tens of feet from the vein. The Gold Hill deposits are a conspicuous exception. There the host rock, the Loomis pluton, is visibly altered, at distances up to half a mile, west of the main axis of mineralization (along Deer Creek) but not to the east (pl. 1). Close to the deposit the mafic minerals of the quartz diorite are completely converted to chlorite, and the feldspar is kaolinized. The halo of alteration can be microscopically detected (green discoloration of

TABLE 7.—Mineral deposits

[All elevations cited below estimated by inspection

Data mostly from literature cited; underlined

Name	Location	Type of deposit	Minerals
Adelia	SW$\frac{1}{4}$SE$\frac{1}{4}$ sec. 16, T. 39 N., R. 26 E. (?).	Vein, 5 ft. wide in slate.	
Alice (probably same as Summit, Palmer Summit, Grand Summit)	SE$\frac{1}{4}$ sec. 30, T. 39 N., R. 26 E.	Summit: Two veins $20 av.; mill recovered $1,538 from 85 tons in 1937. Palmer Summit: 50 tons av. $20 Au. $\frac{1}{2}$- to 6-ft. vein at N. 55° W., nearly vertical dip, in aphanitic greenstone and md. gr. metagabbro. Chlorite alteration near vein.	Free Au in qtz. Py, cpy, gal. Qtz, sid, gal, cpy, py, mala.
American Strategic Minerals	SW$\frac{1}{4}$NE$\frac{1}{4}$ sec. 9, T. 40 N., R. 26 E.	Radioactive black magnetic mineral. Workings in intrusive breccia where malignite extensively diked and brecciated by aplite-alaskite.	
Anaconda	SE$\frac{1}{4}$ sec. 14, T. 39 N., R. 26 E.	4-ft. vein; sample from dump assayed $30 Au, $8 Ag.	Py, cpy, qtz.
Baltimore	Sec. 28, T. 39 N., R. 26 E.	12-in. vein, assaying $10 Au.	Cpy, py.
Bellevue	SE$\frac{1}{4}$NW$\frac{1}{4}$ sec. 4, T. 39 N., R. 26 E.	Vein N. 29° E., dipping 45–66° SE., wall rock is clay slate, vein is 10 to 36 inches wide, 15 in. av. Saw 1-ft. vuggy qtz vein N. 45° W., 50° SW.; also pyritic felsite sill N–S, 50° W., explored by pits.	Arpy, py, cpy, pyrar, steph, native Ag, free Au, and possibly Au-Te. A test shipment of 1,000 lbs. gave $75 per ton in Au and Ag, with Au representing a little over half.

(See footnotes at end of table.)

of Loomis quadrangle

from contour map or by pocket altimeter]

words indicate data added by the authors

Metals	Production	Workings	References[1]	Remarks
Au, Ag.	Several shipments prior to 1902.		6, p. 102.	Not visited.
Au, Ag (Ref. 1, p. 149). Au, Cu, Pb (Ref. 1, p. 146).	1937: $1,900, 1938? $1,000 prior to 1897. 1937, 1939.	Three shafts totaling 420 ft. Five adits totaling 1,200 ft. Developed on 3 levels by adits, stoped to surface.	Summit: 1, p. 149. Grand Summit: 1, p. 146; 7, p. 100. Palmer Summit: 3, p. 106.	
		One adit, one pit.	1, p. 353.	
Au, Ag, Cu.		40-ft. inclined shaft.	1, p. 135 (from 6, p. 102).	Believe shaft to be filled with trash.
Au, Cu.			1, p. 135; 7, p. 101.	Not visited.
Au, Ag, Cu, As, Te, Sb.	$\frac{1}{2}$ ton for $37.50.	Small shaft, 250 ft. development work (7, p. 100).	8, p. 101; 7, p. 100.	

TABLE 7.—Mineral deposits of

Name	Location	Type of deposit	Minerals
Ben Butler	SE¼(?)NE¼ sec. 19, T. 39 N., R. 26 E. [2]	15-in. vein, assay $7.80 Au (1897).	
Black Bear	SE¼NE¼ sec. 36, T. 39 N., R. 25 E.	4-ft. qtz vein along contact of chlorite schist and "serpentine". 3-ft. pay streak. Assays $18 Au, Ag, Cu. (Recon. in 1963 revealed 2 veins at N. 65° W., 75° N.). Country rock is phlebitic grst.	Free Au, py, qtz.
Blackbird	SW½SE¼ sec. 11, T. 39 N., R. 25 E.	Disseminated chromite in partially serpentinized dunite.	Chromite.
Blanche	NW¼NE¼ sec. 10, T. 40 N., R. 26 E.	Vein, with assays of $56, Au–Ag across widths of 42 to 48 in. reported. Vein set N. 50° E., 50° NW.	
Buckeye (see E-9, which is probably coincident with this deposit)	Sec. 26, T. 39 N., R. 26 E.	12-in. vein in quartzite and schist.	Py, cpy, gal, steph.
Bullfrog	S½SW¼ sec. 33, T. 40 N., R. 26 E.	7-ft. qtz vein in quartzite and sericitic schist traceable for 3,000 ft.; 10-ton test yielded $12 Au, $5 Ag (1902).	Py, black sulphide, qtz.
California (probably same as Eagle, Ref. 1, p. 218; Ref. 7, Palmer Mtn. map)	1/10 mile N. of center of sec. 23, T. 40 N., R. 25 E.	6-ft. to 12-ft. vein of vuggy qtz, with rough banding of ore minerals; wall rock is granite.	Cpy, bor, py, sph, gal, mala, lim.

(See footnotes at end of table.)

Loomis quadrangle—Continued

Metals	Production	Workings	References[1]	Remarks
Au.			1, p. 136; 7, p. 101.	
Au, Ag, Cu.	$150,000 prior to 1902, 77 tons in 1947. $113,000 in 1892, (7, p. 98).	2,500 ft. developed by shaft, stopes to surface. 300-ft. shaft.	1, p. 136; 3, p. 106; 7, p. 98.	Ref. 3: qtz vein N. 65° W., 80° NE., 15 to 36 in. at surface, pinching out on 300-ft. level; said to run $24 in Au.
Cr.	None.	None.	1, p. 38	
Au, Ag.		80-ft. and 100-ft. adits; two 10-ft. shafts.	1, p. 136.	
Au, Ag, Pb, Cu, Sb.			3, p. 102.	
Au, Ag.	4,600 lbs. returned $150 in Au, Ag.	Adit, 140-ft. shaft, 160-ft. shaft; subsurface workings.	1, p. 137; 3, p. 101; 7, p. 100.	
Cu, Pb, Zn.		150-ft. adit.	1, p. 63; 3, p. 92; 5, p. 20.	

TABLE 7.—Mineral deposits of

Name	Location	Type of deposit	Minerals
Catherine	Secs. 22 and 23, T. 40 N., R. 25 E.		
Chicago (4 claims)	SE¼ sec. 23, T. 39 N., R. 26 E.	Several veins over 2 ft. wide. Assay $64 Au, $16.40 Ag, $3.60 Pb (1897).	
Chloride Queen	SE¼NE¼ sec. 36, T. 40 N., R. 25 E.	1- to 4-ft. qtz vein in argillite is sparsely mineralized.	Py, free Au.
Cleve	Secs. 4 and 5, T. 38 N., R. 25 E.²ᐟ	15 ft. of low-grade ore, parallel to E. Pluribus.	
Combination	SW¼ sec. 25, T. 39 N., R. 26 E.	4-ft. vein. Ore from dump assayed $21.50 Au, $3 Ag. "Scoriaceous" 1-ft. qtz vein E-W, 90°, pinches out 4-ft. downdip in pit.	Qtz, py, gal, yellow bloom.
Commanding (part of Black Bear-Alice Group)		N. 23° W., E. dip at high angle; vein, 5 to 6 ft. wide poorly defined walls, traceable for several thousand feet.	
Contention	SE. cor. sec. 22, T. 39 N., R. 26 E.	Two veins crossing one another.	
Copper World (includes Copper King)	S½SW¼ sec. 20, T. 39 N., R. 26 E.		Py, cpy, pyrrh, arpy.

(See footnotes at end of table.)

Loomis quadrangle—Continued

Metals	Production	Workings	References[1]	Remarks
Ag, Pb, Zn, Cu.			1, p. 301.	Probably synonym; not shown on map.
Au, Ag, Pb.			1, p. 137; 7, p. 100, 101.	Not visited.
Au.	1936, 1937.	50-ft. inclined shaft, drift.	1, p. 137.	No workings seen in this area. Possibly mis-located.
		Open cut.	7, p. 104.	Not visited.
Au, Ag.		25-ft. shaft, 200-ft. tunnel. Saw pit 6 ft. deep.	1, p. 131; 7, p. 101.	Phyllite wall rock shows drag, indicating left-lateral movement on vein structure.
Cu.		Not developed.	3, p. 106.	
Au.		95-ft. shaft.	1, p. 137; 7, p. 100.	Not visited.
Cu, Au, Ag.		135-ft. inclined shaft, shallow pits.	1, p. 64; 3, p. 104; 7, p. 101.	

TABLE 7.—Mineral deposits of

Name	Location	Type of deposit	Minerals
Copper World Extension (Iron Mask, Iron Master?)	S$\frac{1}{2}$SE$\frac{1}{4}$ sec. 20, N$\frac{1}{2}$NE$\frac{1}{4}$ sec. 29, T. 39 N., R. 26 E. (East of Copper World).	Mineralized zone at N. 85° W., 40° SW. with massive sulfides in en echelon pods. Country rock is altered greenstone. Hanging wall apparently a fault.	Py, cpy, azur, mala, pyrrh, arpy, mag, sph, qtz.
Curlew (Riverview just to south)	Sec. 23, T. 40 N., R. 25 E.	Three parallel veins in granite, assaying $40 to $80 Au (1897).	Gal.
Defiance	NW$\frac{1}{4}$ sec. 4, T. 38 N., R. 26 E. $\frac{2}{}$	3-ft. fissure-fill qtz vein N. 20° W., 35° NE, layering parallel to walls; another vein poorly exposed 200 ft. to NW.	Qtz, gal, py.
Detroit-Windsor	NE$\frac{1}{4}$ sec. 24, T. 38 N., R. 25 E.	14-ft. iron-capped ledge 2-5 percent Cu, some Au.	Fe and Cu, py.
Double Standard Group (includes Copper King, Eclipse, and Mammoth)	Secs. 28 and 33, T. 39 N., R. 25 E.	Iron cap 50 to 75 ft. wide.	Cu and Fe, py.
Eureka	Secs. 17 and 18, T. 40 N., R. 26 E.	Qtz veins 3 to 7 ft. wide. One traceable for 1500 ft. Assays show only trace of Au.	Gal.

(See footnotes at end of table.)

Loomis quadrangle— Continued

Metals	Production	Workings	References[1]	Remarks
Cu (4 percent), Au (1 oz.), Au ($1-50 cents,) W?, Zn?.	3,486 tons 1918-1919 (3.147 percent Cu, 0.42 oz. Ag, 0.03 oz. Au); also production prior to 1911.	300-ft. shaft with cross-cuts on 100-ft., 200-ft., and 300-ft. levels.	3, p. 104; 1, p. 64.	Mine maps.[3]
Au, Pb, Ag.			1, p. 138; 7, (map of Palmer Mountain) and p. 102.	Not visited.
Au, Ag.		Shaft (collapsed), pits, and trenches.	1, p. 138; 7, p. 101.	
Au, Cu.		100-ft. shaft.	1, p. 138; 7, p. 105.	Not visited.
Probably Cu with minor Au, but oxidized surface croppings showed primarily Au (up to $11 to $16), 1897.		10-ft. shaft on Double Standard.	1, p. 64; 7, p. 104.	Not visited.
Pb, Ag, Au.		120-ft. shaft, 110-ft. shaft, 50-ft. shaft, 10-ft. shaft, 30-ft. shaft, two 20-ft. shafts.	1, p. 218.	Not visited.

TABLE 7.—Mineral deposits of

Name	Location	Type of deposit	Minerals
Favorite	SW$\frac{1}{4}$SW$\frac{1}{4}$ sec. 13, T. 40 N., R. 25 E.		
Four Metals Mine	W$\frac{1}{2}$SW$\frac{1}{4}$ sec. 23, T. 40 N., R. 25 E.	Qtz veins av. 4 to 5 ft. wide, (1 to 12 in.) exposed for several thousand feet and depth of 250 ft. Banded sulfides. Scheelite concentrated in vein near walls. Assay est. to average \$8 Pb, Zn, Cu, Ag. On ridge crest east of main adit. Vein set N. 5°–10° W., 25°–30° W., 1/2 in. to 48 in. thick. Offset 6 ft. by fault at N. 30° W., 65° E. At southernmost workings vein 6 ft. thick N. 30° W., 30° SW.	Gal, cpy, py, sph, scheelite, bor, moly, qtz, garnet, epidote.
Gladstone	Sec. 17, T. 39 N., R. 26 E. $\frac{2}{}$	Four parallel ledges, 18-, 14-, 12-, and 13-ft. wide between walls of diorite and porphyry.	
Gold Crown	Sec. 31, T. 39 N., R. 26 E.	10-ft. vein.	Qtz.
"Gold Hill" Group (includes E. Pluribus, and Frankie Girl, etc.	Secs. 4, 5, and 8, T. 38 N., R. 25 E. $\frac{2}{}$	a. Vein at N. 50° E. ? b. Shear zone at N. 55° E., 80° NW., carries pods and veinlets of mineralized qtz; strike slip. c. Qtz vein at N. 65° E., 60°–70° NW., 1 to 2 ft. thick; wall rocks not sheared, except for gouge zone along hanging wall. d. Qtz vein at N. 65° E., 65° NW., about 8 in. thick.	Qtz, gal, py, cpy, mala, azur, sph.

(See footnotes at end of table.)

Loomis quadrangle—Continued

Metals	Production	Workings	References[1/]	Remarks
Pb, Ag, Cu, Au.		850-ft. adit.	1, p. 219; 3, p. 90.	
Pb, Ag, Cu, Zn, W.	20 or more cars of high-grade ore and concentrates 1918-1921. 600 tons milled in 1939.	110-ft. shaft, 150-ft. shaft, 340-ft. adit, 235-ft. adit, 140-ft. adit.	1, p. 219.	See footnote 3.
Au(?).		500 ft. of tunnels and drifts.	1, p. 139; 7, p. 100.	Not visited.
Au.			1, p. 139.	Not visited.
Au, Ag, Cu, Pb, Zn.		Main adit at 3,650 ft. elev., heading S. 65°W., explores vein "b". Numerous other workings. Stopes on vein uphill from main adit caved to surface.	Gold Hill: 1, p. 140; Frankie Girl: 1, p. 139; 7, p. 104; E. Pluribus: 1, p. 64; 7, p. 104.	Country rock (quartz-diorite to granodiorite) intensely altered (chloritized, kaolinized) in vicinity of mineralized area.

TABLE 7.—Mineral deposits of

Name	Location	Type of deposit	Minerals
Gold King	E$\frac{1}{2}$ sec. 11, T. 40 N., R. 26 E.	Graphitic quartz bands in crumpled schists assayed up to $5.75, $2 av.	
Golden Fleece	NE$\frac{1}{4}$ sec. 4, T. 38 N., R. 25 E.	"NE–SW vein," 2$\frac{1}{2}$ to 3 ft.	
Golden Zone	SW$\frac{1}{4}$SW$\frac{1}{4}$ sec. 8, T. 40 N., R. 25 E.	Vein at N. 20° E., 40° NW. in granite. Free Au except where Pb encountered. Ag never over 4 oz. Fault zone at N. 60° E., 45° NNW. crops out west of main adit at about 1,920 ft. elev., contains sheared qtz up to 1 in. thick, pinches to 0. To NE., vein curves to N. 70° E., 80° W. and plunges under talus. To SW. continues with spurs at 1 to $\frac{1}{2}$ foot thick for 100 ft. Wall rocks visibly altered up to 1 ft. from vein.	Py, cpy, Au, gal, (argent.), sph, arpy, qtz, mala, azur.
Grandview (Miller and Redpath)	Secs. 11 and 12, T. 39 N., R. 25 E.	"Several ledges, including one 19$\frac{1}{2}$ to 22 ft. wide, a parallel ledge 14 ft. wide, and a 3-ft. cross-ledge." Portal along fault zone S. 45° W., 80° SE. Fault cuts grst, gneiss, and qtzite.	Barren qtz; vein material on dump; pyritic wall rock.
Hercules	NW$\frac{1}{4}$ sec. 15, T. 38 N., R. 25 E.	Iron cap 80 ft. wide, at E–W, 45° N., traced for over a mile. Surface assays show $2 Au, 5-9 oz. Ag, trace Cu.	Au, Cu, Ag.
Hiawatha (Josie)	NW$\frac{1}{4}$NE$\frac{1}{4}$ sec. 10, T. 39 N., R. 26 E.	Vein 1 to 12 ft. wide, av. 3 to 4 ft., outcropping for about 2,500 ft., dips 10° west at surface, 45° west at depth. Fault offsets vein.	Auriferous gal, cpy, py, sph.

(See footnotes at end of table.)

Loomis quadrangle—Continued

Metals	Production	Workings	References [1]	Remarks
Au.			5, p. 5.	Not visited
Au.		Shallow shafts.		Not visited.
Au, Ag, Pb, Cu, Mo?	Prior to 1911; 1939.	5,000 ft. of tunnels and drifts.	1, p. 140; 3, p. 95; 7, p. 103; 8, p. 95.	
Au, Cu.		Open cross cut, tunnel; shaft 33 ft.	7, p. 103.	
			1, p. 141; 7, p. 104.	Not visited.
Au, Ag, Pb, Zn, Cu.	1938.	Two 80-ft. adits 80 ft. apart, connected by drift near face.	1, p. 141; 3, p. 99; 5, p. 3.	

TABLE 7.—Mineral deposits of

Name	Location	Type of deposit	Minerals
Hoosier?	NW¼NE¼ sec. 26, T. 40 N., R. 25 E.	Qtz vein N. 15° W., 50° W., 1 to 2 ft. wide at 1,500 ft. elev.; N. 35° W., 45° SW., 1 ft. wide at 1,725 ft. elev. (Vein intersects zone of shear at 1,725 ft. elev., N. 60° E., 85° SE. Vein pinches out at shear, but widens on other side. Alteration goes through. Mineralization therefore later than the N. 60° E. shear).	Py.
Horn Silver	SW¼SE¼ sec. 21, T. 40 N., R. 26 E.	Three veins. Most development on one at N. 53° E., 40° NW., 4 in. to 4 ft. wide, av. 18 in.; irregular in strike, dip.	Steph, cerargyrite, proustite, gal, sph, py, cpy, qtz; metallics irregularly distributed throughout vein.
Iron Master	E½ sec. 8, W½ sec. 9, T. 39 N., R. 26 E.	20-ft. to 250-ft. wide, iron cap on ledge.	Iron sulfides.
Ivanhoe	SE¼SW¼ sec. 16, T. 39 N., R. 26 E.	Vein 3½ to 4½ ft. wide. Fine-grained iron-stained qtz vein.	Py, steph, cerargyrite, ruby, silver, qtz; yield from 1,000 ton hand-sorted ore:
Ivanhoe Tunnel	NE¼NE¼ sec. 17, T. 39 N., R. 26 E.		Shipment (lbs) — Ag (oz) — Au (oz): 6,899 / 572 / 1.62; 15,521 / 278 / 1; 25,500 / 326 / 1
John Judge (Leadville)	SW¼SW¼ sec. 19, T. 39 N., R. 26 E.	Vein at N. 60° E., vertical, 2 to 3 ft. wide. Wall rock is chloritic schist.	Gal, py, cpy, bor, pyrar, free Au, qtz, calcite.

(See footnotes at end of table).

Loomis quadrangle——Continued

Metals	Production	Workings	References [1]	Remarks
Au.		Tunnel at 1,500-ft. elev., tunnel at 1,725 ft.	7, p. 102	"Ledge 47 ft. wide"— Ref. 7; may be a massive quartzite bed.
Ag, Cu, Au, Sb, Pb, Zn.	Two carloads av. $62 per ton ($2.50 was in Cu, Au, remainder in Ag) 1909.	750-ft. adit, 100-ft. raise, 100-ft. shaft.	3, p. 100; 1, p. 304.	
Au, Ag, up to $6; Cu (a "little").			7, p. 101.	Not visited.
Ag, Au, Sb.	1900.	65-ft. inclined shaft (500 ft. according to Ref. 7); 70-ft. drift at 120-ft. level. Ivanhoe Tunnel 4,400 ft.	1, p. 304; 7, p. 99; 3, p. 102.	Shipment cited was from oxidized and enriched zone near surface.
Au, Cu, Pb, Ag.	1937, 15 ton, 1938, 1939.	2,500 ft. of adits and shafts.	1, p. 143; 3, p. 104; 7, p. 100.	Vein is slickensided; alteration more intense along south wall. Galena in separate cluster of crystals.

TABLE 7.—Mineral deposits of

Name	Location	Type of deposit	Minerals
Julia (part of San Francisco Group)	SW$\frac{1}{4}$ sec. 24, T. 40 N., R. 25 E.	7-ft. qtz vein.	Gal.
Kaaba Texas (originally Caaba.)	Center of NE$\frac{1}{4}$ sec. 23, T. 40 N., R. 25 E.	Qtz vein, banded. From 1943-1946 average was $4.03 per ton; 1.2 percent Pb, 0.5 percent Zn, 0.1 percent Cu, 2.25 oz. Ag. Vein is 6 to 12 ft. wide at N. 2° W., 45° W. exposed for 600 ft. on strike and opened for 150 ft. on dip (1917).	Gal, cpy, sph, moly, py, scheelite, qtz, calcite.
Kalamazoo	NE$\frac{1}{4}$ sec. 33, T. 39 N., R. 26 E.	50-ft. wide zone with sulfides.	Qtz and native Au.
Kimberly	Near center of SW$\frac{1}{4}$ sec. 11, T. 39 N., R. 26 E.	En echelon lenses of qtz in hanging wall over contact (possibly fault), at N. 46° W., 52° SW., between igneous (footwall) and metamorphic rocks.	Gal, py, cpy, qtz.
King Solomon	NE$\frac{1}{4}$SW$\frac{1}{4}$ sec. 24, T. 40 N., R. 25 E.	Banded vuggy qtz vein at N. 5° E., 50° W., 7 ft. wide at outcrop, pinching at depth (slightly) of 10-15 ft. (limit of visibility). Wall rock thin lam. Slate and qtzite.	Gal, cpy, py.
Lakeview (probably synonym of Empire, and also Palmer Lake)	SW$\frac{1}{4}$SW$\frac{1}{4}$ sec. 32, T. 40 N., R. 26 E. (practically on township line).	Vein 2 to 3$\frac{1}{2}$ ft. thick, at N. 60° W., 35° NE.; 3 ft. thick at portal; widens locally to 7 ft.; thins on strike to 1 ft., 30 ft. to E. in massive siltstone.	Py, cpy, bor, qtz, arpy.

(See footnotes at end of table).

Loomis quadrangle—Continued

Metals	Production	Workings	References [1]	Remarks
Pb, Ag, Au.		150-ft. shaft, 118-ft. adit.	1, p. 220; 7, p. 101.	Not visited.
Pb, Ag, Zn, Cu, Au, W, Mo?	1915, 1918, 1920, 1929, 1943-49 (95,000 tons), 1950.	300-ft. inclined shaft; 1,100 ft. of drifts on four levels.	1, p. 220.	See footnote 3.
Cu, Au.			1, p. 67; 7, p. 101.	Not visited.
Pb, Au, Ag.	Considerable hand-sorted ore of $40 grade (1917). Ref. 5.	140-ft. inclined shaft; drifts at 60-ft., 80-ft., 100-ft. levels.	1, p. 220; 3, p. 97; 5, p. 3.	
Pb, Cu.	1924.	Inclined shaft at least 30-40 ft. deep; another adit 50 feet lower. (Both inaccessible.)	1, p. 220.	
Au, Cu.		30-ft. inclined shaft.	1, p. 69; 3, p. 101; 1, p. 138 (Empire); 7, p. 101 (Empire).	

TABLE 7.—Mineral deposits of

Name	Location	Type of deposit	Minerals
Lancashire Lass	SE$\frac{1}{4}$ sec. 27, T. 39 N., R. 26 E. (adjoins Rainbow).		
Little Chopaka (Defense, probably synonym of Worthington)	SE$\frac{1}{4}$SE$\frac{1}{4}$ sec. 16, T. 40 N., R. 25 E.	Disseminated chromite; also in thin seams in serpentinite.	Chromite.
Little Chopaka (Six Eagles)	SW$\frac{1}{4}$SE$\frac{1}{4}$ sec. 14, T. 40 N., R. 25 E.	1-ft. to 3-ft. qtz vein in fault zone in granite, N. 2° E., 45° W.	Py, cpy, sph, gal, qtz.
Little Falls	NE$\frac{1}{4}$ sec. 15, T. 38 N., R. 25 E.	2-ft. qtz vein.	
Lone Pine (Submarine)	NW$\frac{1}{4}$NW$\frac{1}{4}$ sec. 3, T. 40 N., R. 26 E.	Blanket qtz vein, at N. 50° E., flat to ± 10°, 1 to 10 ft., 6 ft. av. Traceable on strike for 1,000 ft.	Argentiferous gal, cpy, py, cov, bor, qtz.
Mammoth (adjoins Eclipse and Copper King)	(See Double Standard)		

(See footnotes at end of table).

Loomis quadrangle—Continued

Metals	Production	Workings	References [1]	Remarks
Au.			1, p. 143; 7, p. 100.	Not visited.
Cr.			1, p. 38, 39.	
Pb, Ag, Au.		2,000-ft. adit, 200-ft. shaft. Tunnel, 500 feet below shaft, cut seam of gouge instead of vein.	1, p. 221; 3, p. 90-91.	
Au.		50-ft. shaft.	1, p. 143; 7, p. 104.	Not visited.
Ag, Pb, Cu, Au.		300-ft. tunnel at N. 50° W., 1,000-ft. tunnel 110 ft. below upper, which failed to intersect ore body.	2, p. 243-244; 1, p. 306.	Vein shattered by post-mineral movement.

TABLE 7.—Mineral deposits of

Name	Location	Type of deposit	Minerals
McGrath	NW$\frac{1}{2}$SW$\frac{1}{4}$ sec. 13(?), T. 40 N., R. 25 E. ($\frac{1}{2}$ mile south of Nighthawk).	Vein 2 to 3 ft. wide.	
Midas No. 1	SE$\frac{1}{4}$ sec. 24, T. 39 N., R. 25 E.	Shear zone in gabbro with disseminated? Cu minerals. Numerous barren qtz veins.	Cpy, py, mala.
Minnie	SE$\frac{1}{4}$ sec. 14, T. 39 N., R. 26 E. (eastern extension of Anaconda).	2-ft. vein.	
Mountain Sheep	NE$\frac{1}{4}$NW$\frac{1}{2}$ sec. 28, T. 40 N., R. 25 E.	On extension of Ruby mineralization fault; vein traceable for 3,000 ft. along outcrop. Fault N. 65° W., 35° SW. locally mineralized; qtz up to 6 ft. thick. Wall rock visibly altered up to 4 ft. from fault; crops out at about 1,950 ft. elev. Second vein crops out at about 2,090 ft. elev. at N. 80° W., 45° S.; $\frac{1}{2}$ to 1$\frac{1}{2}$ ft. thick.	Qtz, mala.
Nighthawk	NE$\frac{1}{4}$SW$\frac{1}{4}$ sec. 13, T. 40 N., R. 25 E.	Irreg. mineralized qtz bodies along margin of brecciated zone in granite; zone is 100 feet wide at one place.	Py, gal, qtz.
Ninety-Two	Sec. 21, T. 39 N., R. 26 E.		

(See footnotes at end of table).

Loomis quadrangle—Continued

Metals	Production	Workings	References [1]	Remarks
Zn, Ag.		Several thousand feet of adits.	1, p. 365.	
Cu.		30-ft. drift.	1, p. 67.	
Au.		Shaft.	1, p, 144.	
Ag, Au (trace).	A few cars prior to 1911.	Three adits totaling 2,000 ft. Main adit at about 1,450 ft. elev.; no vein material on dump.	1, p. 307; 3, p. 95.	
Pb, Ag.	Some.	1,770-ft. adit.	1, p. 221; 3, p. 89-90.	
Au.		160-ft. tunnel.	1, p. 145; 7, p. 100.	Not visited.

TABLE 7.—Mineral deposits of

Name	Location	Type of deposit	Minerals
Number One	NW$\frac{1}{4}$NW$\frac{1}{4}$ sec. 23, T. 40 N., R. 25 E.	Qtz vein 3 to 12 ft. wide in fault zone, N-S, 31° W.; vein traceable for 4,000 ft.	Gal, qtz.
Oro Fino	W$\frac{1}{2}$ sec. 23, T. 38 N., R. 25 E.[2/]	4-ft. vein, N. 35° E. 90°; qtz filled fault zone along felsite dike.	Py, qtz.
Palmer Mtn. Tunnel	NE$\frac{1}{4}$NE$\frac{1}{4}$ sec. 1, T. 38 N., R. 25 E.	Failed to intersect Black Bear vein at depth. Encountered thin sparsely mineralized veins.	Mala, cpy, gal.
Peerless (probably includes older Wyandotte group, Ref. 7, p. 102, and Palmer Mountain map.)	NW$\frac{1}{4}$SE$\frac{1}{4}$ sec. 22, T. 40 N., R. 25 E.	Evidently a zone of contact metamorphism adjacent to granite to north, with scattered showing of Cu.	Cpy and contact meta. Minerals and mag.
Pinnacle	NE$\frac{1}{4}$SE$\frac{1}{4}$ sec. 24, T. 39 N., R. 25 E.	Pinnacle vein N. 60° E., vertical, and Bunker Hill vein, E-W. At intersect of these two, qtz was 4 to 10 ft. wide and averaged $11 across face (pre-1910). Country rock is diabase but highly altered.	Au, py, cpy, sph, qtz, calcite.
Prize	SE$\frac{1}{4}$SW$\frac{1}{4}$ sec. 25, T. 40 N., R. 25 E.	Qtz vein at N. 75° E., 60° SE. at 3,130 ft. elev., N. 85° E. at 3,100 ft. elev. (2$\frac{1}{2}$ ft. thick). N. 85° E., 45° S. at 3,040 ft. elev., 1$\frac{1}{2}$ ft. thick. At 2,990 ft. level attitude as above, 2 ft. thick. Vein is banded in lower part.	Qtz, arpy, gal, py, cpy, mala, azur, and iron oxides.

(See footnotes at end of table.)

Loomis quadrangle—Continued

Metals	Production	Workings	References [1]	Remarks
Pb, Ag, (Au, Cu, minor).		200-ft. adit, 60-ft. shaft.	1, p. 221; 3, p. 91-92.	
Au.		Minor shafts, adits, trenches, and pits.	1, p. 145.	
Au, Ag (Cu, Pb, minor).		6,610-ft. tunnel, drifts aggregating 2,000 ft.; diamond drilled 800 ft. beyond face of tunnel.	1, p. 146; 3, p. 106-107; 7, p. 98.	
Cu.		400-ft. tunnel.	3, p. 93; 1, p. 69.	
Au, Cu, Pb, Zn, Ag.	Ref. 3 credits it with perhaps $150,000 (p. 76) in Au.	2,000 ft. of workings in three adits.	3, p. 105; 1, p. 146.	
Pb, Ag, Au, Cu.	1906, 1913.	Three tunnels on vein, at about 3,100 ft., 3,040 ft., 2,990 ft. elev. Above upper tunnel is holed-through stope at about 3,130 ft. elev.	1, p. 222; 3, p. 93.	Some brecciated vein quartz.

TABLE 7.—Mineral deposits of

Name	Location	Type of deposit	Minerals
Rainbow	Center of NE¼ sec. 22, T. 39 N., R. 26 E.	Pinching and swelling qtz lenses in lms, qtzite, and schist.	Py, arpy, cpy, gal.
Red Jacket	SE¼ sec. 14, T. 38 N., R. 25 E.	3-ft. vein.	
Rich Bar	SW¼NW¼ sec. 11, T. 40 N., R. 26 E.	Vein at N. 40° W., 35°-40° NE., stringers to 6 ft. wide. Erratic values.	Cpy, sph, py, steph, gal, argentite, qtz.
Rich Bar Placer	SW¼NW¼ sec. 11, T. 40 N., R. 26 E.		
Riverview	Sec. 23, T. 40 N., R. 25 E.	4-ft. ledge.	
Ruby (includes Beggars Choice) Formerly (continued on next page)	NE¼SE¼ sec. 28, T. 40 N., R. 25 E.	Mineralization along N. 45° W., 45° SW. fault; partial fissure fill and gouge. Mineralized zone av. 3 feet thick.	Py, cpy, arpy, gal, sph, proustite, pyrar, argentite, Au, mala, azur, lim.

(See footnotes at end of table.)

Loomis quadrangle—Continued

Metals	Production	Workings	References	Remarks
Ag, Au, Cu, Pb.	High grade as high as 200 oz. Ag and several dollars Au. (Ref. 5).	Three adits. 150-ft. tunnel on main ledge with 65-ft. winze and 312-ft. cross-cut (Ref. 7). 316-ft. cross-cut on another claim inter-sects vein at 110-ft. depth.	3, p. 102; 1, p. 147; 7, p. 100; 5, p. 3.	
Au.	2 tons of hand-picked ore at $40 each, in 1892.		1, p. 147; 7, p. 104.	Not visited.
Cu, Zn, Ag, Pb.		150-ft. shaft, with drifts from 50 ft. and 150 ft. levels.	1, p. 70; 3, p. 95-96.	
Au.			3, p. 95.	Ref. infers that Rich Bar Lode is at site of old Rich Bar Placer, worked as far back as 1859.
		100-ft. shaft.	1, p. 147; 7, p. 102.	Not visited
Ag, Pb, Au, Cu, Zn.	1915-1920, $25,000 (31 cars); 1922-1923; 1939.	5,000-ft. adit S. 55° W.; 950-ft. drifts at adit level,	1, p. 309; 3, p. 94-95; 2, p. 237-240.	

TABLE 7.—Mineral deposits of

Name	Location	Type of deposit	Minerals
(continued from preceding page) "Rush"(?) (Refs. 1, p. 148; 7, p. 103, and map of Palmer Mountain)		At about 1,550-ft. elev., mineralized fault at N. 25° E., 50° NW. Ore minerals in breccia of wall rock and qtz, about 1 to $\frac{1}{2}$ ft. thick. Wall rock sheered and altered up to 2 ft. from main structure. Other faults at N. 65° W., 40° N.; E-W., 35° N.; N. 60° E., 75° S.; N. 5° E., 50° W., all with evidence of mineralization.	Qtz, calcite, gouge (hornblende).
Saint	Sec. 2, T. 40 N., R. 26 E.		
San Francisco Group (includes Ellemeham, California, California Cross Course, Pontiac, Kelly, and Julia)	Secs. 18 and 19, T. 40 N., R. 26 E., and secs. 13 and 24, T. 40 N., R. 25 E.[2/]		Gal.
Second Prize (same as Gearhart?)	NE$\frac{1}{4}$?SW$\frac{1}{4}$ sec. 19, T. 39 N., R. 26 E.	Second Prize claim: Qtz vein N. 54° E., 60° NW., 3 to 4 ft. wide. Columbia Claim: N. 15° E., 60° NW., on vein which pinches out at 35 ft.	Auriferous py, arpy.
Security	Sec. 36, T. 39 N., R. 25 E.	Shear zone with vein 70 ft. long, and 0 to 3 ft. wide. Sparsely mineralized.	Py, cpy, gal, sph, qtz.
Silent Friend	SE$\frac{1}{4}$ sec. 10, T. 39 N., R. 26 E.	Vein.	Py, qtz.

(See footnotes at end of table.)

Loomis quadrangle—Continued

Metals	Production	Workings	Reference[1]	Remarks
		also 210-ft. winze. Main adit about 1,200 ft. elev.		
Au, Ag, As.			1, p. 148.	Not visited.
Pb, Au.		60-ft. tunnel, 100-ft. tunnel, 90-ft. shaft.	1, p. 148; 7, p. 101.	Not visited.
Au.		Tunnels 70 ft. and 125 ft. (Second Prize). Shaft, 35 ft. (Columbia).	1, p. 148; 3, p. 104.	
Cu, Pb, Zn, Au.		620-ft. adit, 200-ft. adit.	1, p. 70.	Not visited.
Ag, Au.		30-ft. shaft.		Not visited.

TABLE 7.—Mineral deposits of

Name	Location	Type of deposit	Minerals
Spokane (American Rand)	NE¼SE¼ sec. 10, NW¼SW¼ sec. 11, T. 39 N., R. 26 E.	Qtz vein 15 in. wide.	Py, gal, cpy, moly, Au.
Standard	Sec. 10, T. 39 N., R. 26 E.	Three parallel ledges. Evidently the largest averages 4 ft. wide.	
Summit	NE¼NW¼ sec. 23, T. 40 N., R. 25 E.	Qtz vein 3½ to 4 ft. wide exposed for 500 feet along strike at N. 10° W., 40° SW.	Py, gal, (cpy, sph, moly, minor) qtz, sericite.
Tenderfoot (Gold Crown and Gold Eagle claims)	S½NE¼ sec. 5, T. 39 N., R. 25 E.	Veins a few inches to 4 ft. thick in granite.	Py, cpy, qtz.
Treasury Group	Sec. 26?, T. 38 N., R. 26 E.	24-ft. ledge of rose qtz.	
Trinity	NE¼ sec. 28, T. 40 N., R. 26 E.	Irregular lenses of somewhat mineralized qtz in calcareous shale.	
Triune (includes Caroline, Jessie, and Occident)	E½NE¼ sec. 10, W½NW¼ sec. 11, T. 39 N., R. 26 E.	4 qtz veins, stringers to 10 ft. thick. N-S, 20–40° E. (very irreg. in thickness and dip). Principal production from this vein. At lower level, 4 veins striking west of north, dipping west, and av. less than 18 in. thick. Wall rock clayey, qtzite to 75-ft. level, granite below.	Py, gal, cpy, moly, Au, mala, azur, (gal is auriferous). Assays: 4,000 tons of tailings yielded 0.01 oz Au, 0.3 oz Ag. Sample of 3 ledges on south side of gulch gave 6.32 oz Au, 0.4 oz Ag, 5 ft. of ore on N. side gave 0.03 oz Au, 0.05 oz Ag.
Utica	W½ sec. 23, T. 38 N., R. 25 E. 2/	5-ft. qtz ledge in NE-SW porphyry dike (more Cu than Gold Hill ledges).	Hematite, qtz.

(See footnotes at end of table.)

Loomis quadrangle—Continued

Metals	Production	Workings	References[1]	Remarks
Au, Ag, Pb, Cu, Mo.	Prior to 1900; 1916-1918; 1935-1938.	100-ft. inclined shaft with drifts.	1, p. 149; 3, p. 98; 7, p. 99.	
Au, low grade.		Various small openings.	7, p. 100.	Not visited.
Ag, minor Pb, Au, Cu.		70-ft. inclined shaft.	1, p. 310; 2, p. 231; 3, p. 91; 7, p. 104.	
Au, Ag, Cu.		60-ft. shaft.	1, p. 139.	Not visited.
Au.		80-ft. shaft, trenches, 200-ft. tunnel.	7, p. 105.	Not visited.
Au, Ag.		40-ft. shaft, 200-ft. adit.	1, p. 150; 3, p. 99.	Not visited.
Au, Ag, Pb.	More than $300,000 prior to 1930; 1939.	140-ft. shaft, 2,000-ft. drifts.	1, p. 150; 3, p. 97-98; 7, p. 99; 5, p. 2.	
Au, Ag, Cu.		60-ft. shaft.	1, p. 150; 7, p. 104.	Not visited. (Refs. list as NW$\frac{1}{4}$ sec. 26).

TABLE 7.—Mineral deposits of

Name	Location	Type of deposit	Minerals
Wannacut Lake	Sec. 31, T. 40 N., R. 26 E.		
War Eagle	SE¼ sec. 36, T. 39 N., R. 25 E.	5-ft. qtz vein.	Au.
Wehe (see E-3, for analyses of vein and wall rock)	NE¼ sec. 26, T. 39 N., R. 26 E.	Vein, ladder-type in shear. N-S 90°, wall rock meta-siltstone and metagabbro.	Qtz, gal, sph, CuS, cpy, calcite, mala (rare).

All data from this

Name	Location	Type of deposit	Minerals	Workings
E-1	Sec. 35, T. 39 N., R. 26 E.	E-W qtz vein, dip?, 6 ft. horizontal width in one pit.	Sulfide?, py, lim, qtz.	Two pits on vein.
E-2	Sec. 26, T. 39 N., R. 26 E.	Vein N. 10° E., 70° E., apparently lenticular (reef type).	Qtz, calcite, lim, siderite.	Pit.
E-3 [5/]	NE¼ sec. 26, T. 39 N., R. 26 E.			
E-4	Sec. 26, T. 39 N., R. 26 E.	Vein filling in fault zone N. 10° E., about 90°.	Qtz, gal, py.	Pit 12 ft. deep.

(See footnotes at end of table.)

Loomis quadrangle—Continued

Metals	Production	Workings	References[1]	Remarks
Au.		At least 400 feet.	1, p. 150.	Not visited.
Au, Cu, Ag.	$150,000 prior to 1902.	100-ft. adit, two 70-ft. drifts.	1, p. 150; 7, p. 98.	Not visited.
Au, Ag.		Small pit; shaft, 50 ft. to water.	1, p. 150; 7, p. 101.	

study by the authors

	Semiquantitative spectrographic analyses[4]							Remarks
Sample	Element							
V=vein C=country rock	Ag	$\frac{Bi}{Cd}$	Cu	Pb	Sb	Zn	Other	
V	0	0	0.0003	0	0	0	---	Country rock meta-sediments.
V	0	0	0.0015	0	0	0		Wall rocks metasediments and meta-diabase.
C	0	0	0.002	0.002	0	0	---	
C	0.00007	0	0.007	0.0015	0	0		
V	0.3	$\frac{0}{0.3}$	1.0	10.0	3.0	10.0	As 5.0	See Wehe deposit.
C	0.0003	0	0.007	0.015	0	0	---	
C	0.00015	0	0.007	0.01	0	0	---	
V	0.15	$\frac{0}{11}$	0.2	7.0	0.5	2.0		
C	0.00007	0	0.015	0.005	0	0	---	
C	0.00007	0	0.02	0.003	0	0		
C	0.0003	$\frac{0.00015}{0}$	0.007	0.015	0	0		

TABLE 7.—Mineral deposits of

Name	Location	Type of deposit	Minerals	Workings
E-5	Sec. 25, T. 39 N., R. 26 E.	En echelon qtz saddle-type veins, up to 5 ft. thick, tapering to a few inches in 10 to 20 ft., concordant with wall rock.	Qtz, gal, calcite (vein is drusy).	Three adits, caved.
E-6	Sec. 25, T. 39 N., R. 26 E.	?Small qtz pods and veinlets on dump; wall rock mi, slate, phyllite.	Qtz, py, lim, calcite.	One adit, caved; dump volume about 11,000 cu. ft.; several pits.
E-7	Sec. 26, T. 39 N., R. 26 E.	En echelon qtz, veins at N. 15° W., 80° W. to 70° E., pinch and swell from 1 to $2\frac{1}{2}$ ft. down dip and pinch out on strike.	Qtz, calcite, cpy, mala, azur; noted cpy in wall rock.	One shaft, 15 ft. to water; two pits; several trenches.
E-8	Sec. 25, T. 39 N., R. 26 E.	Qtz veins. West vein is 2 ft. wide at surface, pinches out 10 ft. down. East vein is mineralized 4 to 12 in. wide; fault zone, N. 15° E., 65° W., country rocks are ppy w/qtz.	Qtz, lim, gal, py, mala.	Shaft 15 to 20 ft. deep.
E-9[6/]	Sec. 26, T. 39 N., R. 26 E.	Qtz vein in shear zone in mi at N. 35° E., 80-90° E., and 4 to 10 inches wide in upper adit. Flat-abutting qtz vein has drag suggesting a west down movement. Anastomosing qtz pods in 5 ft. wide zone at N. 5-15° E., vertical to 80° W. Vein at N. 15° E., 85° W. Barren qtz vein 6 feet thick at N. 45° E., 50° SE., drusy.	Qtz, py, mala, cpy, gal, calcite, arpy, sericite.	Two adits on vein separated 40 ft. vertically; 10,000 cu. ft. dump at lower adit; 20-ft. shaft.

(See footnotes at end of table.)

Loomis quadrangle—Continued

| Sample | Semiquantitative spectrographic analyses [4] | | | | | | | Remarks |
V=vein C=country rock	Ag	Bi/Cd	Cu	Pb	Sb	Zn	Other	
V	0.015	0 / 0.03	0.003	0.7	0	1.5	---	
C	0.0007	0	0.01	0.005	0	0		
V	0.00015	0	0.03	0.003	0	0	---	
C	<0.00007	0	0.02	0.002	0	0		
V	<0.00007	0	0.0015	0.005	0	0		
C	0.00007	0	0.2	0.003	0	0		
C	0.0015	0	0.007	0.015	0	0		Note two parallel possibly post-mineral faults N. 15° E., 65° NW., with vertical slickensides.
C	0	0	0.003	0.002	0	0		
V	0.5	0.015 / 0.01	0.5	20.0	0.7	0.2		
C	0.0005	0	0.007	0.02	0	0		
C	0	0	0.007	0.003	0	0		
C	0.0001	0	0.01	0.002	0	0		
C	0.0001	0	0.01	0.002	0	0		
C	0.00007	0	0.003	0.003	0	0		
C	0	0	0.015	0.003	0	0		
C	0	0	0.007	0	0	0		
C	0.0003	0	0.007	0.003	0	0		
C	0.00007	0	0.005	0	0	0		
C	<0.00007	0	0.003	0	0	0	As 0.3	
C	0.0005	0	0.015	0.002	0	0		
C	0.0001	0	0.001	0.002	0	0	As 0.2	
V	0.00015	0	0.0015	0.01	0	0		
C	<0.00007	0	0.001	0.0015	0	0		
V	0.005	0	0.0015	0.3	0	0.07	Au 0.005	
V	0.1	0.003 / 0.03	2.0	15.0	0.07	0.15		

TABLE 7.--Mineral deposits of

Name	Location	Type of deposit	Minerals	Workings
E-10a	Sec. 25, T. 39 N., R. 26 E.	Qtz pods? on dump; wall rock phy, and mi.	Py, gal, qtz.	Adit N. 60° E., dump about 20,000 cu. ft.
E-10b	· · · · do · · · ·	Qtz abundant on dump; probably gash filling in phyllite?	Py, in vuggy qtz (scoreaceous), gal, calcite.	Adit S. 75° E., more than 50 ft. long.
E-10c	· · · · do · · · ·	Qtz vein 6 to 10 ft. thick, pod type, but traceable to E-10b?	Qtz, gal, py.	Shallow pit.
E-11	Sec. 25, T. 39 N., R. 26 E.	Saddle-type qtz vein in crest of minor anticlinal fold in meta; 3 ft. thick, pinches out at ends; 25 ft. long.	Qtz, (drusy) py, gal.	Pit.
E-13	Sec. 36, T. 39 N., R. 26 E.	Qtz pod conformable, 6 ft. thick, 8 ft. wide.	Drusy qtz, py, lim with excellent cubic boxworks.	Two pits.
E-14	Sec. 23, T. 39 N., R. 26 E.	Vein at N. 40° E., 50° SE., ½ to 1 ft. thick, mostly calcite except thin qtz seams with sulfides along footwall. Mineralized post vein fault up to ½ ft. thick; qtz, at main shaft. Also vein in fault zone at N. 40° E., 75° W., about 200 ft. to NE. of shaft.	Calcite, qtz, gal.	Shaft with incline to west at bottom; 20 ft. to water. 10,000 to 20,000 cu. ft. dump.

(See footnotes at end of table.)

Loomis quadrangle—Continued

Semiquantitative spectrographic analyses [4/]								Remarks
Sample	Element							
V=vein C=country rock	Ag	Bi / Cd	Cu	Pb	Sb	Zn	Other	
V	0.003	0.005 / 0.1	0.015	0.1	0	3.0		Yellow efflorescence on walls, associated with "scoriaceous" qtz.
V	0.01	0.007 / 0	0.05	0.15	0	0	As 2.0	
V	0.02	0.015 / 0	0.03	0.3	0	0	As 20.0	
C	0.00015	0	0.015	0.005	0	0		
								Post-mineral fault N. 15° E., 80° SE.; slight drag indicates vertical movement; qtz brecciated.
C	<0.00007	0	0.01	0.002	0	0		Post mineral-fault N. 15° E., 80° S.E., slickensides plunge 15° N.
C	0	0	0.003	0.003	0	0		
C	<0.00007	0	0.01	0	0	0		
V	0.0007	0	0.2	0.02	0	0.1		

TABLE 7.—Mineral deposits of

Name	Location	Type of deposit	Minerals	Workings
E-15	Sec. 24, T. 39 N., R. 26 E.	Qtz vein system; fissure fill along fault. N. 80° E., 80° S. with qtz bands along south wall, horizontal qtz veins joining and pinching out to north, stoping width was 3 to 4 ft. Wall rock is mi.	Qtz, gal, py, sph, CuS, mala.	Adit N. 80° E., more than 50 ft. long; inclined shaft 20 ft. to water; stope holed through to surface above adit.
E-16	Sec. 15, T. 38 N., R. 26 E.	Contact metamorphic? with grano-diorite? intruding lms.	Qtz, arpy.	Inclined shaft N. 85° W., 75°; 20 ft. to water; several pits.
E-17	Sec. 36, T. 39 N., R. 26 E.	Py disseminated in lms; also some replacement gal? Most of the pits are on 1- to 1½-ft. phyllite interbeds in lms, except winze is on fault zone at N. 30° E., dipping steeply NW.; drag indicates W-down.	Py, gal(?).	Shaft 25 ft. deep; adit with winze 25 ft. to water; drift off adit 15 ft. from portal.
L-111	Sec. 11, T. 39 N., R. 26 E.	E-W, 20° N. fault zone mineralized with bull qtz.	Qtz.	
L-112	Sec. 11, T. 39 N., R. 26 E.	1-ft. qtz vein at N. 30° W., vertical.	Qtz, yellow and white gypsum? precipitation on walls. Sulfides (arpy abundant; gal, rare).	10-ft. adit.

(See footnotes at end of table.)

Loomis quadrangle—Continued

Semiquantitative spectrographic analyses [4]								Remarks
Sample	Element							
V=vein C=country rock	Ag	$\dfrac{Bi}{Cd}$	Cu	Pb	Sb	Zn	Other	
V	0.7	$\dfrac{0.002}{0.3}$	3.0	30.0	5.0	15.0	Au 0.007	Slickensided py indicates post-mineral faulting. Sample selected of considerably higher grade than average.
V C C C C	0.0003 <0.00007 0.00007 <0.00007 <0.00007	0 0 0 0 0	0.003 0.01 0.01 0.015 0.005	0.01 0.003 0 0.003 0.002	0 0 0 0 0	0 0 0.07 0.05 0	As 0.3	
V C	0.05 0.00007	0 0	0.03 0.0015	0.5 0.003	0.1 0	0.2 0		

No semiquantitative spectrographic analyses
run on samples L-111 and L-112.

TABLE 7.—Mineral deposits of

Name	Location	Type of deposit	Minerals	Workings
L-166	Sec. 4, T. 39 N., R. 26 E.	$\frac{1}{2}$ ft. wide qtz vein at N. 40° W., 80° SW. in mi.	Vuggy qtz with gal.	8-ft. pit.
L-184	Sec. 32, T. 40 N., R. 26 E.	Qtz vein 1 ft. thick, N. 65° E., 85° SE., wall rock is Bullfrog Mountain Formation.	Qtz.	
L-185	Sec. 32, T. 40 N., R. 26 E.	Qtz vein $\frac{1}{2}$ ft. at N. 35° W., 65° SW., irregular, shootlike, pinches out at depth.	Vuggy qtz, with abundant gal.	6-ft. pit.
L-188	Sec. 10, T. 39 N., R. 26 E.	Qtz vein N. 70° W., 60° N., 5 ft. wide in center of shoot, pinching to 1 ft. at upper and lower edges of exposure (15 ft. high). Second vein is 1 ft. wide at N. 55° W. dipping steeply to SW.	Qtz, py, gal?	Small pits.
L-192	Sec. 28, T. 40 N., R. 26 E.	8- to 12-in. wide qtz vein at N. 25° W., 40° NE.	Qtz, cpy, green stain.	Six pits and trenches.
L-199	Sec. 29, T. 40 N., R. 26 E.	Two anastomosing qtz veins pinch and swell from 1 to 18 inches at about N. 30° E., 55° NW. in mi.	Qtz, mala, cpy.	Pit.
L-217	Sec. 17, T. 40 N., R. 26 E.	2-ft. qtz shoot at N. 40° E., 85° SE.	Qtz.	

(See footnotes at end of table).

Loomis quadrangle—Continued

Semiquantitative spectrographic analyses [4]								Remarks
Sample	Element							
V=vein C=country rock	Ag	$\dfrac{Bi}{Cd}$	Cu	Pb	Sb	Zn	Other	
								Vein is cut by faults N. 30° E., 40° NW.
No semiquantitative spectrographic analyses run on samples L-166 through L-217.								

TABLE 7.—Mineral deposits of

Name	Location	Type of deposit	Minerals	Workings
L-219	Sec. 17, T. 40 N., R. 26 E.	1- to 1½-ft. vein at N. 35° E., 45° NW., irregular at least 100 ft. long.	Qtz, gal, py.	Pit.
L-220	Sec. 21, T. 40 N., R. 26 E.	Vein set at N-S at N. 30° E., dipping vertically to 70° W., 1 to ½ ft. to several inches thick; continuous for several tens of feet on strike.	Qtz.	Pit.
L-222	Sec. 16, T. 40 N., R. 26 E.	1- to ½-ft. vein (fissure fill) at N. 30° W., 80° NE.	Cpy, mala, qtz.	Shaft 30 ft. to water; 100 ft. lower on hill is tunnel (open) with qtz on dump.
L-230	Sec. 24, T. 40 N., R. 25 E.	About 2- to ½-ft. vein at N. 45° E., 80° NW.	Qtz, arpy, gal.	Two pits on vein.
L-231	Sec. 24, T. 40 N., R. 25 E.	Qtz vein (fissure fill) with layering parallel to walls of vein (suggestion of comb structure); 2½ to 3 ft. thick, N. 15° W., 60° W.	Qtz, gal, mala.	Pit and short-tunnel at 2,700 ft. elev., 20-ft. shaft at 2,780 ft. elev.
L-255	Sec. 16, T. 40 N., R. 26 E.	Qtz vein and zone of veinlets 6 in. to 1 ft. thick at E-W, 40° N.	Qtz, py.	Incline, 25 ft. to water.
L-301	Sec. 3, T. 40 N., R. 26 E.	Vein set at N. 70° W., 25° N.	Qtz.	

(See footnotes at end of table).

Loomis quadrangle—Continued

Semiquantitative spectrographic analyses [4]									Remarks
Sample	Element								
V=vein C=country rock	Ag	Bi / Cd	Cu	Pb	Sb	Zn	Other		
No semiquantitative spectrographic analyses run on samples L-219 through L-301.									Vein cut off by sill of Cenozoic volcanic rock trending N. 60° E., 75° W., 10 to 20 ft. thick. Vein qtz locally slickensided.
									Veins cut out by N. 10° W. vertical fault; vein locally shattered.

TABLE 7.—Mineral deposits of

Name	Location	Type of deposit	Minerals	Workings
L-302	Sec. 14, T. 40 N., R. 25 E.	$\frac{1}{2}$- to $2\frac{1}{2}$-ft. vein at N. 5° E. 35° E. (about $2\frac{1}{2}$ ft. thick along nearly continuous exposure for 500 ft. north of locality).	Qtz, with drusy gal, lim, mala.	Pits on vein.
L-334	Sec. 19, T. 38 N., R. 26 E.	Several narrow veins at N. 50° W., vertical.	Qtz.	
L-362	Sec. 25, T. 39 N., R. 25 E.	4-ft. zone of shear with qtz stringers. Qtz not well mineralized but wall rock and gouge locally high-grade pyrite; N. 20° W., 70° NE.	Cu-stain, py.	Shaft.
L-703	SE$\frac{1}{4}$ sec. 13, T. 40 N., R. 25 E.	About 25 ft. brecciated, vuggy qtz vein N. 10° E., 20° W.	Qtz, gal, py.	Pit on vein.

1/ References cited:

1. Huntting, M. T., 1956, Metallic minerals— Part 2 of Inventory of Washington minerals: Washington Div. Mines and Geology Bull. 37, v. 1, 428 p.

2. Patty, E. N., 1921, The metal mines of Washington: Washington Geol. Survey Bull. 23, 366 p.

3. Umpleby, J. B., 1911b, Geology and ore deposits of the Oroville-Nighthawk mining district: Washington Geol. Survey Bull. 5, part 2, p. 53-107, 110-111.

4. Spedden, H. R., Jr., 1939, Mine of the American Rand Corporation near Oroville: Univ. of Washington B. S. thesis, 65 p.

5. Handy, F. M., [1916?], An investigation of the mineral deposits of northern Okanogan County: Washington State College Dept. of Geology Bull. 100, 27 p.

6. Bethune, G. A., 1892, Mines and minerals of Washington: Washington State Geologist, 2d Annual Report, 1891, 187 p.

7. Hodges, L. K., 1897, Mining in the Pacific Northwest: The Seattle Post-Intelligencer, Seattle, Wash., 192 p.

8. Smith, G. O.; Calkins, F. C., 1904, A geological reconnaissance across the Cascade Range near the forty-ninth parallel: U.S. Geol. Survey Bull. 235, 103 p.

Loomis quadrangle—Continued

Semiquantitative spectrographic analyses [4]								Remarks
Sample	Element							
V=vein C=country rock	Ag	$\dfrac{Bi}{Cd}$	Cu	Pb	Sb	Zn	Other	
			No semiquantitative spectrographic analyses run on samples L-302 through L-703.					

[2] Locality changed to agree with official records on file with U.S. Department of the Interior, Bureau of Land Management.

[3] Mine map(s) on file at Washington Division of Mines and Geology, Department of Natural Resources, Olympia, Washington.

[4] Analyst, Chris Heropoulos. Results are reported in percent to the nearest number in the series 1, 0.7, 0.5, 0.3, 0.2, 0.15, and 0.1, etc.; which represent approximate midpoints of interval data on a geometric scale. The assigned interval for semiquantitative results will include the quantitative value about 30 percent of the time.

[5] Locality probably same as Wehe deposit.

[6] Probably equivalent to Buckeye deposit.

Abbreviations used in table: Minerals—arpy, arsenopyrite; azur, azurite; bor, bornite; cpy, chalcopyrite; CuS, copper sulfide; cov, covellite; gal, galena; lim, limonite; mag, magnetite; mala, malachite; moly, molybdenite; pyrar, pyrargarite; py, pyrite; pyrrh, pyrrhotite; qtz, quartz; sid, siderite; sph, sphalerite; steph, stephanite. Rocks—grst, greenstone; lms, limestone; mi, mafic intrusive rock; meta, metamorphic rock; phy, phyllite; ppy, porphyritic or porphyry; qtzite, quartzite.

the normally brown biotite) past the limit shown on plate 1.

An area of alteration of completely different character was mapped near American Butte within the Similkameen composite pluton (sec. 1, T. 40 N., R. 25 E.). Here the normally massive granodiorite is altered to a tough sericitic quartz-albite greisen, laced with quartz pods and veinlets. The quartz pods contain minor proportions of calcite (magnesian?), irregular to subpolygonal (pseudomorphous?) clots of chlorite, and pyrite. Thin rinds of similarly altered rock were noted adjacent to a few of the quartz veins in and near Similkameen pluton.

Oxidation of primary copper and silver-lead ore minerals has not progressed far enough to offer much opportunity for secondary enrichment, as primary sulfides usually can be found close to the surface of the outcrop. The Ivanhoe mine is evidently an exception. According to Hodges (1897, p. 99), the upper portions of the vein were considerably enriched in silver oxides and native silver. Gold in the upper portions of some of the veins was free-milling and somewhat enriched (Handy, 1916[?], p. 5; Spedden, 1939, p. 22) due to oxidation of associated sulfides.

The age of the quartz veins can be roughly estimated as mid-Cretaceous to early Tertiary. The older limit is set by their occurrence in the youngest plutonic rocks of the quadrangle (Anderson Creek and Similkameen plutons). No mineralization was observed in the Eocene sedimentary rocks, and, in addition, several of the veins are cut by dacite dikes correlated on the basis of appearance with Eocene dacite plugs; thus, the quartz veins must be older than Eocene. No evidence bearing on the age of the massive sulfide deposits was found.

The outlook for further production from most of the vein deposits is not particularly encouraging. Apparently the readily accessible and high-grade portions of the productive veins have already been mined. The remaining mineralized rock is probably too low grade or not available in large enough quantities to warrant exploitation at present prices and with present technology. On the other hand, the possibility that larger or higher grade undiscovered deposits exist cannot be ruled out. Areas which appear to us to offer the most potential are: (1) Gold Hill and the valley of Deer Creek; (2)

Copper World Extension deposits; and (3) the altered area within the Similkameen pluton north of Nighthawk.

Magmatic Deposits

Chrome.—Chromite as disseminations and thin veinlets has been found in the serpentinized dunite-peridotite rock of Little Chopaka Mountain and probably also occurs in similar rocks on Chopaka and Grandview Mountains. A reconnaissance of these areas suggests that the concentration of chromite is too low to constitute a potential resource. Deposits reported on Chopaka Mountain in the adjacent Horseshoe Basin quadrangle were not investigated.

Iron.—The pyroxenite associated with the eastern border of the Similkameen composite pluton contains disseminated magnetite. Chemical analyses of two specimens gave 7.3 and 10.2 percent Fe_2O_3, respectively, and 10.0 and 8.9 percent FeO, respectively. If these specimens are indicative of the iron content of the body as a whole, the grade is not high.

Placer Deposits

The gold-bearing gravels along the lower course of the Similkameen River were among the first deposits in the area to attract the attention of prospectors. The volume of auriferous gravel was not large, and the deposits were essentially worked out in the latter part of the nineteenth century, although desultory sluicing and "sniping" continues to the present day. The presence of platinum was reported by Huntting (1956); the source of the platinum is possibly iron-bearing pyroxenite.

NONMETALLIC DEPOSITS

Limestone

Some of the limestone beds of the Spectacle Formation are of sufficient size and purity to be potential economic sources of high-calcium limestone. The limestones and associated rocks were studied by Crosby (1949), and his map was later published by Mills (1962), who reviewed the pertinent features of the deposits and published analyses of specimens from several localities. Mills concluded (1962, p. 224) that, ". . . the chances are very good of there being many millions of tons of high-calcium limestone."

Tonasket Lime Products, Inc. operates a crushing and grinding plant near Spectacle Lake and markets a high-calcium limestone quarried from one of the beds of the Spectacle Formation (sec. 13, T. 38 N., R. 26 E.).

Gypsite

A considerable amount of gypsite has been concentrated as an evaporite in a playa at Lenton Flat near the northwestern corner of the quadrangle in sec. 3, T. 40 N., R. 25 E. The playa covers about 80 acres east of the Similkameen River and is about 150 feet above it. The deposit was not studied by us, but was described by Bennett (1962, p. 124, 125), who included in his report two chemical analyses (1962, table 39), listed below, of material obtained from a test hole at the west end of the deposit. He states, ". . . assuming

that similar material extends throughout the lake bed to a depth of at least 5 feet, there are available approximately 1,000,000 tons of intermixed gypsite, basic magnesium carbonate, calcium carbonate, and other constituents including clay."

Analyses of salts from Lenton Flat[1]

	No. 77 (0-6 ft.)	No. 78 (6-16 ft.)
CaO	29.7	21.6
MgO	8.8	3.8
Al_2O_3	1.0	8.4
Fe_2O_3	0.5	1.7
Na_2O	. . .	1.4
SiO_2	6.4	29.0
CO_2	10.3	6.6
SO_3	37.4	22.8
Sulfide	trace	trace
Cl	0.2	1.3
Ignition loss	16.3	9.1
Total	100.3	99.1

[1] Constituents in percent of total weight.

REFERENCES CITED

Adams, R. W., 1962, Geology of the Cayuse Mountain-Horse Springs Coulee area, Okanogan County, Washington: Univ. of Washington M.S. thesis, 41 p.

Armstrong, J. E., 1949, Fort St. James map-area, Cassiar and Coast districts, British Columbia: Geol. Survey of Canada Memoir 252, 210 p.

Bateman, P. C.; Clark, L. D.; Huber, N. K.; Moore, J. G.; Rinehart, C. D., 1963, The Sierra Nevada batholith; a synthesis of recent work across the central part: U.S. Geol. Survey Prof. Paper 414-D, p. D1-D46.

Bennett, W. A. G., 1962, Saline lake deposits in Washington: Washington Div. Mines and Geology Bull. 49, 129 p.

Bethune, G. A., 1892, Mines and minerals of Washington: Washington State Geologist, 2d Annual Report, 1891, 187 p.

Bostock, H. S., 1940, Keremeos, Similkameen district, British Columbia: Canada Geol. Survey Map 341A, scale 1:63,360.

Bostock, H. S., 1941a, Okanagan Falls, Similkameen and Osoyoos districts, British Columbia: Canada Geol. Survey Map 627A, scale 1:63,360.

Bostock, H. S., 1941b, Olalla, Similkameen, Osoyoos, and Kamloops districts, British Columbia: Canada Geol. Survey Map 628A, scale 1:63,360.

Campbell, C. D., 1939, The Kruger alkaline syenites of southern British Columbia: American Journal of Science, v. 237, no. 8, p. 527-549.

Crosby, J. W., 3d, 1949, Limestones in the Anarchist Series, Okanogan County, Washington: Washington State College M.S. thesis, 15 p.

Daly, R. A., 1906, The Okanagan composite batholith of the Cascade Mountain system: Geol. Soc. America Bull., v. 17, p. 329-376.

Daly, R. A., 1912, Geology of the North American Cordillera at the forty-ninth parallel: Canada Geol. Survey Memoir 38, parts 1-3, 857 p.

Dawson, G. M., 1890, Report on a portion of the West Kootanie district, B. C.: Canada Geol. Survey Annual Report 4, 66 p.

Dawson, W. L., 1898, Glacial phenomena in Okanogan County, Washington: American Geologist, v. 22, p. 203-217.

Engel, A. E. J.; Engel, C. G., 1962, Hornblendes formed during progressive metamorphism of amphibolites, northwest Adirondack Mountains, New York: Geol. Soc. America Bull., v. 73, no. 12, p. 1499-1514.

Flint, R. F., 1935, Glacial features of the southern Okanogan region: Geol. Soc. America Bull., v. 46, no. 2, p. 169-194.

Folk, R. L., 1959, Practical petrographic classification of limestones: American Assoc. Petroleum Geologists Bull., v. 43, no. 1, p. 1-38.

Fryxell, Roald, 1960, Problems in glacial chronology of northern Washington [abstract]: Geol. Soc. America Bull., v. 71, no. 12, part 2, p. 2060-2061.

Goldsmith, Richard, 1952, Petrology of the Tiffany-Conconully area, Okanogan County, Washington: Univ. of Washington Ph. D. thesis, 356 p.

Greenwood, H. J., 1962, Metamorphic reactions involving two volatile components: Carnegie Inst. Washington Yearbook 61, 1961-62, p. 82-85.

Handy, F. M., [1916?], An investigation of the mineral deposits of northern Okanogan County: Washington State College Dept. of Geology Bull. 100, 27 p.

Harker, R. I.; Tuttle, O. F., 1955a, The thermal dissociation of calcite, dolomite and magnesite — Part 1 of Studies in the system CaO-MgO-CO_2: American Jour. Science, v. 253, no. 4, p. 209-224.

Harker, R. I.; Tuttle, O. F., 1955b, Limits of solid solution along the binary join $CaCO_3$-$MgCO_3$ — Part 2 of Studies in the system CaO-MgO-CO_2: American Jour. Science v. 253, no. 5, p. 274-282.

REFERENCES CITED—Continued

Hess, H. H., 1939, Island arcs, gravity anomalies and serpentinite intrusions — A contribution to the ophiolite problem: Internat. Geol. Congress, 17th, Moscow 1937, Report, v. 2, p. 263-283.

Hibbard, M. J., 1962, Geology and petrology of crystalline rocks of the Toats Coulee Creek region, Okanogan County, Washington: Univ. of Washington Ph. D. thesis, 96 p.

Hibbard, M. J., 1971, Evolution of a plutonic complex, Okanogan Range, Washington: Geol. Soc. America Bull., v. 82, no. 11, p. 3013-3047.

Hodges, L. K., 1897, Mining in the Pacific Northwest: The Seattle Post-Intelligencer, Seattle, Wash., 192 p.

Huber, N. K.; Rinehart, C. D., 1966, Some relationships between the refractive index of fused glass beads and the petrologic affinity of volcanic rock suites: Geol. Soc. America Bull., v. 77, no. 1, p. 101-109.

Huntting, M. T., 1955, Gold in Washington: Washington Div. Mines and Geology Bull. 42, 158 p.

Huntting, M. T., 1956, Metallic minerals — Part 2 of Inventory of Washington minerals: Washington Div. Mines and Geology Bull. 37, v. 1, 428 p.

Huntting, M. T.; Bennett, W. A. G.; Livingston, V. E., Jr.; Moen, W. S., 1961, Geologic map of Washington: Washington Div. Mines and Geology, scale 1:500,000.

Johannsen, Albert, 1931, A descriptive petrography of the igneous rocks — Volume 1, Introduction, textures, classifications, and glossary; Volume 2, The quartz-bearing rocks: Univ. of Chicago Press, Chicago, Illinois, v. 1, 267 p.; v. 2, 428 p.

Jones, A. G., 1959, Vernon map-area, British Columbia: Canada Geol. Survey Memoir 296, 186 p.

Laniz, R. V.; Stevens, R. E.; Norman, M. B., 1964, Staining of plagioclase feldspar and other minerals with F. D. and C. Red No. 2: U.S. Geol. Survey Prof. Paper 501-B, p. B152-B153.

Lawson, A. C., 1896, On malignite, a family of basic plutonic orthoclase rocks rich in alkalies and lime, intrusive in the Coutchiching schists of Poohbah Lake [Ontario]: Univ. of California Dept. of Geology Bull., v. 1, no. 12, p. 337-362.

Little, H. W., 1957, Kettle River (east half), Similkameen, Kootenay, and Osoyoos districts, British Columbia: Canada Geol. Survey Map 6-1957, scale 1:253,440.

Little, H. W., 1961, Geology, Kettle River (west half), British Columbia: Canada Geol. Survey Map 15-1961, scale 1:253,440.

Lounsbury, R. W., 1951, Petrology of the Nighthawk-Oroville area, Washington: Stanford Univ. Ph. D. thesis, 104 p.

Mathews, W. H., 1964, Potassium-argon age determinations of Cenozoic volcanic rocks from British Columbia: Geol. Soc. America Bull., v. 75, no. 5, p. 465-468.

McIntyre, A. W., 1907, Copper deposits of Washington: American Mining Congress, 9th Annual Session, Report of Proceedings, p. 238-250.

Menzer, F. J., Jr., 1964, Geology of the crystalline rocks west of Okanogan, Washington: Univ. of Washington Ph. D. thesis, 64 p.

Mills, J. W., 1962, High-calcium limestones of eastern Washington: Washington Div. Mines and Geology Bull. 48, 268 p.

Misch, Peter, 1951, Large thrusts in northern Cascades of Washington [abstract]: Geol. Soc. America Bull., v. 62, no. 12, part 2, p. 1508-1509.

Misch, Peter, 1966, Tectonic evolution of the northern Cascades of Washington State: Canadian Institute of Mining and Metallurgy Special Volume 8, p. 101-148.

Miyashiro, Akiho, 1961, Evolution of metamorphic belts: Jour. Petrology, v. 2, no. 3, p. 277-311.

Nasmith, Hugh, 1962, Late glacial history and surficial deposits of the Okanogan Valley, British Columbia: British Columbia Dept. Mines and Petroleum Resources Bull. 46, 46 p.

REFERENCES CITED—Continued

Noble, J. A.; Taylor, H. P., Jr., 1960, Correlation of the ultramafic complexes of southeastern Alaska with those of other parts of North America and the world: Internat. Geol. Congress, 21st, Copenhagen, 1960, Report, part 13, p. 188-197.

Nockolds, S. R., 1954, Average chemical compositions of some igneous rocks: Geol. Soc. America Bull., v. 65, no. 10, p. 1007-1032.

Pardee, J. T., 1918, Geology and mineral deposits of the Colville Indian Reservation, Washington: U.S. Geol. Survey Bull. 677, 186 p.

Patty, E. N., 1921, The metal mines of Washington: Washington Geol. Survey Bull. 23, 366 p.

Peacock, M. A., 1931, Classification of igneous rock series: Jour. Geology, v. 39, no. 1, p. 54-67.

Pearson, R. C., 1967, Geologic map of the Bodie Mountain quadrangle, Ferry and Okanogan Counties, Washington: U.S. Geol. Survey Geologic Quadrangle Map GQ-636, scale 1:62,500.

Pelton, H. A., 1957, Geology of the Loomis-Blue Lake area, Okanogan County, Washington: Univ. of Washington M.S. thesis, 92 p.

Rice, H. M. A., 1947, Geology and mineral deposits of the Princeton map-area, British Columbia: Canada Geol. Survey Memoir 243, 136 p.

Shapiro, Leonard; Brannock, W. W., 1956, Rapid analysis of silicate rocks: U.S. Geol. Survey Bull. 1036-C, p. 19-56.

Smith, G. O.; Calkins, F. C., 1904, A geological reconnaissance across the Cascade Range near the forty-ninth parallel: U.S. Geol. Survey Bull. 235, 103 p.

Snook, J. R., 1965, Metamorphic and structural history of "Colville batholith" gneisses, north-central Washington: Geol. Soc. America Bull., v. 76, no. 7, p. 759-776.

Spedden, H. R., Jr., 1939, Mine of the American Rand Corporation near Oroville: Univ. of Washington B.S. thesis, 65 p.

Thompson, M. L.; Wheeler, H. E.; Danner, W. R., 1950, Middle and Upper Permian fusulinids of Washington and British Columbia: Cushman Foundation for Foraminiferal Research Contributions, v. 1, parts 3-4, no. 8, p. 46-63.

Umpleby, J. B., 1911a, Geology and ore deposits of the Myers Creek mining district: Washington Geol. Survey Bull. 5, part 1, p. 1-52, 109-110.

Umpleby, J. B., 1911b, Geology and ore deposits of the Oroville-Nighthawk mining district: Washington Geol. Survey Bull. 5, part 2, p. 53-107, 110-111.

Waters, A. C., 1937, Erosion surfaces in the Wenatchee-Chelan district, Washington [abstract]: Geol. Soc. America Proceedings 1936, p. 319-320.

Waters, A. C., 1939, Resurrected erosion surface in central Washington: Geol. Soc. America Bull., v. 50, no. 4, p. 635-657 [with discussion by Bailey Willis on p. 658-659].

Waters, A. C.; Krauskopf, Konrad, 1941, Protoclastic border of the Colville batholith: Geol. Soc. America Bull., v. 52, no. 9, p. 1355-1417.

White, W. H., 1959, Cordilleran tectonics in British Columbia: American Assoc. Petroleum Geologists Bull., v. 43, no. 1, p. 60-100.

Willis, Bailey, 1887, Changes in river courses in Washington Territory due to glaciation: U.S. Geol. Survey Bull. 40, 10 p.

Winkler, H. G. F., 1965, Petrogenesis of metamorphic rocks: Springer-Verlag, New York City, 220 p.

Wolfe, J. A., 1968, Paleogene biostratigraphy of nonmarine rocks in King County, Washington: U.S. Geol. Survey Prof. Paper 571, 33 p.

Wright, A. E.; Bowes, D. R., 1963, Classification of volcanic breccias — a discussion: Geol. Soc. America Bull., v. 74, no. 1, p. 79-86.

Yates, R. G.; Becraft, G. E.; Campbell, A. B.; Pearson, R. C., 1966, Tectonic framework of northeastern Washington, northern Idaho, and northwestern Montana: Canadian Institute of Mining and Metallurgy Special Volume 8, p. 47-59.

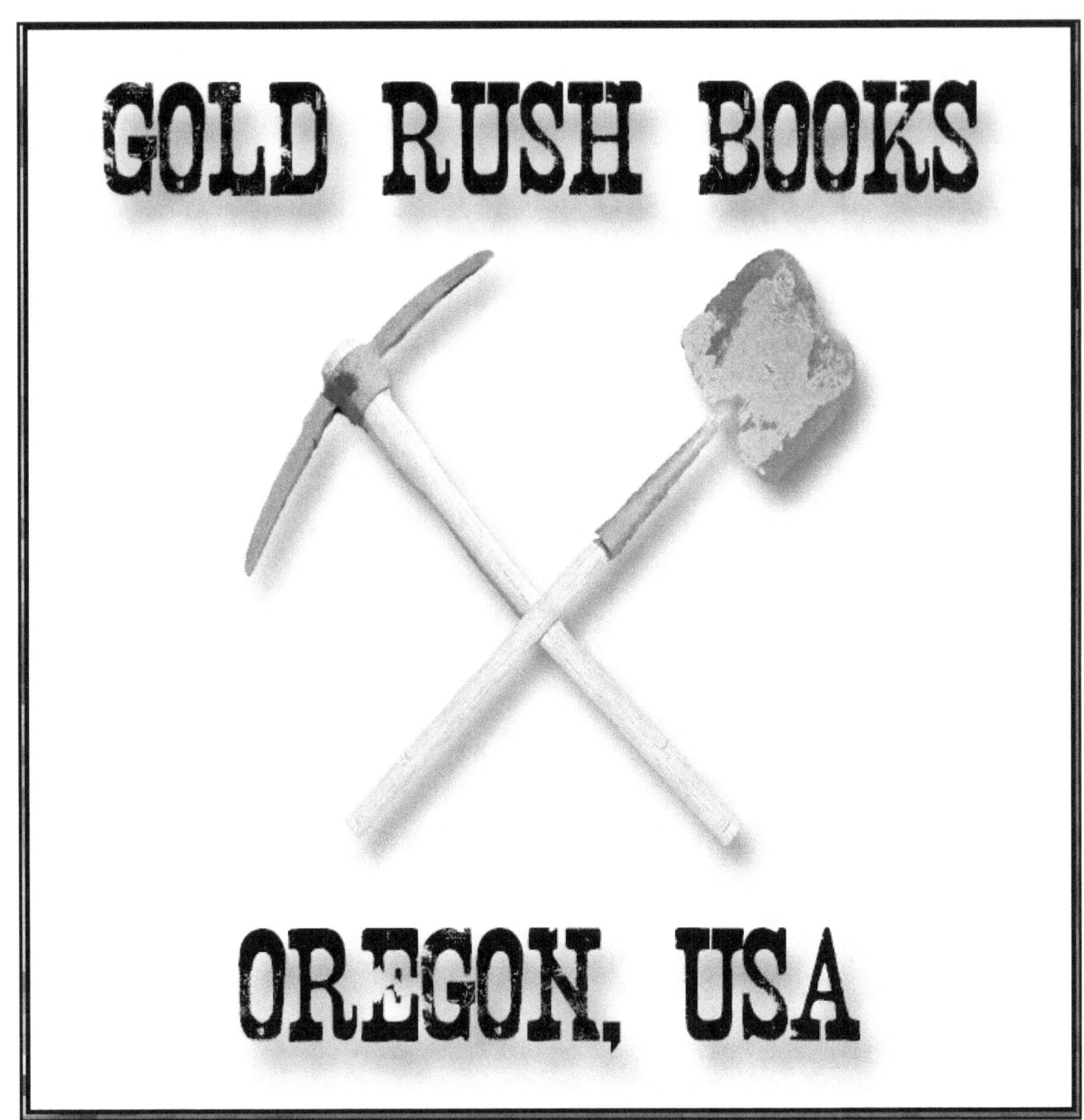

www.GoldMiningBooks.com

Books On Mining

Visit: www.goldminingbooks.com to order your copies or ask your favorite book seller to offer them.

Mining Books by Kerby Jackson

<u>Gold Dust: Stories From Oregon's Mining Years</u> - Oregon mining historian and prospector, Kerby Jackson, brings you a treasure trove of seventeen stories on Southern Oregon's rich history of gold prospecting, the prospectors and their discoveries, and the breathtaking areas they settled in and made homes. 5" X 8", 98 ppgs. Retail Price: $11.99

<u>The Golden Trail: More Stories From Oregon's Mining Years</u> - In his follow-up to "Gold Dust: Stories of Oregon's Mining Years", this time around, Jackson brings us twelve tales from Oregon's Gold Rush, including the story about the first gold strike on Canyon Creek in Grant County, about the old timers who found gold by the pail full at the Victor Mine near Galice, how Iradel Bray discovered a rich ledge of gold on the Coquille River during the height of the Rogue River War, a tale of two elderly miners on the hunt for a lost mine in the Cascade Mountains, details about the discovery of the famous Armstrong Nugget and others. 5" X 8", 70 ppgs. Retail Price: $10.99

Oregon Mining Books

<u>Geology and Mineral Resources of Josephine County, Oregon</u> - Unavailable since the 1970's, this important publication was originally compiled by the Oregon Department of Geology and Mineral Industries and includes important details on the economic geology and mineral resources of this important mining area in South Western Oregon. Included are notes on the history, geology and development of important mines, as well as insights into the mining of gold, copper, nickel, limestone, chromium and other minerals found in large quantities in Josephine County, Oregon. 8.5" X 11", 54 ppgs. Retail Price: $9.99

<u>Mines and Prospects of the Mount Reuben Mining District</u> - Unavailable since 1947, this important publication was originally compiled by geologist Elton Youngberg of the Oregon Department of Geology and Mineral Industries and includes detailed descriptions, histories and the geology of the Mount Reuben Mining District in Josephine County, Oregon. Included are notes on the history, geology, development and assay statistics, as well as underground maps of all the major mines and prospects in the vicinity of this much neglected mining district. 8.5" X 11", 48 ppgs. Retail Price: $9.99

<u>The Granite Mining District</u> - Notes on the history, geology and development of important mines in the well known Granite Mining District which is located in Grant County, Oregon. Some of the mines discussed include the Ajax, Blue Ribbon, Buffalo, Continental, Cougar-Independence, Magnolia, New York, Standard and the Tillicum. Also included are many rare maps pertaining to the mines in the area. 8.5" X 11", 48 ppgs. Retail Price: $9.99

<u>Ore Deposits of the Takilma and Waldo Mining Districts of Josephine County, Oregon</u> - The Waldo and Takilma mining districts are most notable for the fact that the earliest large scale mining of placer gold and copper in Oregon took place in these two areas. Included are details about some of the earliest large gold mines in the state such as the Llano de Oro, High Gravel, Cameron, Platerica, Deep Gravel and others, as well as copper mines such as the famous Queen of Bronze mine, the Waldo, Lily and Cowboy mines. This volume also includes six maps and 20 original illustrations. 8.5" X 11", 74 ppgs. Retail Price: $9.99

<u>Metal Mines of Douglas, Coos and Curry Counties, Oregon</u> - Oregon mining historian Kerby Jackson introduces us to a classic work on Oregon's mining history in this important re-issue of Bulletin 14C Volume 1, otherwise known as the Douglas, Coos & Curry Counties, Oregon Metal Mines Handbook. Unavailable since 1940, this important publication was originally compiled by the Oregon Department of Geology and Mineral Industries includes detailed descriptions, histories and the geology of over 250 metallic mineral mines and prospects in this rugged area of South West Oregon. 8.5" X 11", 158 ppgs. Retail Price: $19.99

Metal Mines of Jackson County, Oregon - Unavailable since 1943, this important publication was originally compiled by the Oregon Department of Geology and Mineral Industries includes detailed descriptions, histories and the geology of over 450 metallic mineral mines and prospects in Jackson County, Oregon. Included are such famous gold mining areas as Gold Hill, Jacksonville, Sterling and the Upper Applegate. **8.5" X 11", 220 ppgs. Retail Price: $24.99**

Metal Mines of Josephine County, Oregon - Oregon mining historian Kerby Jackson introduces us to a classic work on Oregon's mining history in this important re-issue of Bulletin 14C, otherwise known as the Josephine County, Oregon Metal Mines Handbook. Unavailable since 1952, this important publication was originally compiled by the Oregon Department of Geology and Mineral Industries includes detailed descriptions, histories and the geology of over 500 metallic mineral mines and prospects in Josephine County, Oregon. **8.5" X 11", 250 ppgs. Retail Price: $24.99**

Metal Mines of North East Oregon - Oregon mining historian Kerby Jackson introduces us to a classic work on Oregon's mining history in this important re-issue of Bulletin 14A and 14B, otherwise known as the North East Oregon Metal Mines Handbook. Unavailable since 1941, this important publication was originally compiled by the Oregon Department of Geology and Mineral Industries and includes detailed descriptions, histories and the geology of over 750 metallic mineral mines and prospects in North Eastern Oregon. **8.5" X 11", 310 ppgs. Retail Price: $29.99**

Metal Mines of North West Oregon - Oregon mining historian Kerby Jackson introduces us to a classic work on Oregon's mining history in this important re-issue of Bulletin 14D, otherwise known as the North West Oregon Metal Mines Handbook. Unavailable since 1951, this important publication was originally compiled by the Oregon Department of Geology and Mineral Industries and includes detailed descriptions, histories and the geology of over 250 metallic mineral mines and prospects in North Western Oregon. **8.5" X 11", 182 ppgs. Retail Price: $19.99**

Mines and Prospects of Oregon - Mining historian Kerby Jackson introduces us to a classic mining work by the Oregon Bureau of Mines in this important re-issue of The Handbook of Mines and Prospects of Oregon. Unavailable since 1916, this publication includes important insights into hundreds of gold, silver, copper, coal, limestone and other mines that operated in the State of Oregon around the turn of the 19th Century. Included are not only geological details on early mines throughout Oregon, but also insights into their history, production, locations and in some cases, also included are rare maps of their underground workings. **8.5" X 11", 314 ppgs. Retail Price: $24.99**

Lode Gold of the Klamath Mountains of Northern California and South West Oregon
(See California Mining Books)

Mineral Resources of South West Oregon - Unavailable since 1914, this publication includes important insights into dozens of mines that once operated in South West Oregon, including the famous gold fields of Josephine and Jackson Counties, as well as the Coal Mines of Coos County. Included are not only geological details on early mines throughout South West Oregon, but also insights into their history, production and locations. **8.5" X 11", 154 ppgs. Retail Price: $11.99**

Chromite Mining in The Klamath Mountains of California and Oregon
(See California Mining Books)

Southern Oregon Mineral Wealth - Unavailable since 1904, this rare publication provides a unique snapshot into the mines that were operating in the area at the time. Included are not only geological details on early mines throughout South West Oregon, but also insights into their history, production and locations. Some of the mining areas include Grave Creek, Greenback, Wolf Creek, Jump Off Joe Creek, Granite Hill, Galice, Mount Reuben, Gold Hill, Galls Creek, Kane Creek, Sardine Creek, Birdseye Creek, Evans Creek, Foots Creek, Jacksonville, Ashland, the Applegate River, Waldo, Kerby and the Illinois River, Althouse and Sucker Creek, as well as insights into local copper mining and other topics. **8.5" X 11", 64 ppgs. Retail Price: $8.99**

Geology and Ore Deposits of the Takilma and Waldo Mining Districts - Unavailable since the 1933, this publication was originally compiled by the United States Geological Survey and includes details on gold and copper mining in the Takilma and Waldo Districts of Josephine County, Oregon. The Waldo and Takilma mining districts are most notable for the fact that the earliest large scale mining of placer gold and copper in Oregon took place in these two areas. Included in this report are details about some of the earliest large gold mines in the state such as the Llano de Oro, High Gravel, Cameron, Platerica, Deep Gravel and others, as well as copper mines such as the famous Queen of Bronze mine, the Waldo, Lily and Cowboy mines. In addition to geological examinations, insights are also provided into the production, day to day operations and early histories of these mines, as well as calculations of known mineral reserves in the area. This volume also includes six maps and 20 original illustrations. **8.5" X 11", 74 ppgs. Retail Price: $9.99**

Gold Mines of Oregon - Oregon mining historian Kerby Jackson introduces us to a classic work on Oregon's mining history in this important re-issue of Bulletin 61, otherwise known as "Gold and Silver In Oregon". Unavailable since 1968, this important publication was originally compiled by geologists Howard C. Brooks and Len Ramp of the Oregon Department of Geology and Mineral Industries and includes detailed descriptions, histories and the geology of over 450 gold mines Oregon. Included are notes on the history, geology and gold production statistics of all the major mining areas in Oregon including the Klamath Mountains, the Blue Mountains and the North Cascades. While gold is where you find it, as every miner knows, the path to success is to prospect for gold where it was previously found. 8.5" X 11", 344 ppgs. Retail Price: $24.99

Mines and Mineral Resources of Curry County Oregon - Originally published in 1916, this important publication on Oregon Mining has not been available for nearly a century. Included are rare insights into the history, production and locations of dozens of gold mines in Curry County, Oregon, as well as detailed information on important Oregon mining districts in that area such as those at Agness, Bald Face Creek, Mule Creek, Boulder Creek, China Diggings, Collier Creek, Elk River, Gold Beach, Rock Creek, Sixes River and elsewhere. Particular attention is especially paid to the famous beach gold deposits of this portion of the Oregon Coast. 8.5" X 11", 140 ppgs. Retail Price: $11.99

Chromite Mining in South West Oregon - Originally published in 1961, this important publication on Oregon Mining has not been available for nearly a century. Included are rare insights into the history, production and locations of nearly 300 chromite mines in South Western Oregon. 8.5" X 11", 184 ppgs. Retail Price: $14.99

Mineral Resources of Douglas County Oregon - Originally published in 1972, this important publication on Oregon Mining has not been available for nearly forty years. Included are rare insights into the geology, history, production and locations of numerous gold mines and other mining properties in Douglas County, Oregon. 8.5" X 11", 124 ppgs. Retail Price: $11.99

Mineral Resources of Coos County Oregon - Originally published in 1972, this important publication on Oregon Mining has not been available for nearly forty years. Included are rare insights into the geology, history, production and locations of numerous gold mines and other mining properties in Coos County, Oregon. 8.5" X 11", 100 ppgs. Retail Price: $11.99

Mineral Resources of Lane County Oregon - Originally published in 1938, this important publication on Oregon Mining has not been available for nearly seventy five years. Included are extremely rare insights into the geology and mines of Lane County, Oregon, in particular in the Bohemia, Blue River, Oakridge, Black Butte and Winberry Mining Districts. 8.5" X 11", 82 ppgs. Retail Price: $9.99

Mineral Resources of the Upper Chetco River of Oregon: Including the Kalmiopsis Wilderness - Originally published in 1975, this important publication on Oregon Mining has not been available for nearly forty years. Withdrawn under the 1872 Mining Act since 1984, real insight into the minerals resources and mines of the Upper Chetco River has long been unavailable due to the remoteness of the area. Despite this, the decades of battle between property owners and environmental extremists over the last private mining inholding in the area has continued to pique the interest of those interested in mining and other forms of natural resource use. Gold mining began in the area in the 1850's and has a rich history in this geographic area, even if the facts surrounding it are little known. Included are twenty two rare photographs, as well as insights into the Becca and Morning Mine, the Emmly Mine (also known as Emily Camp), the Frazier Mine, the Golden Dream or Higgins Mine, Hustis Mine, Peck Mine and others. 8.5" X 11", 64 ppgs. Retail Price: $8.99

Gold Dredging in Oregon - Originally published in 1939, this important publication on Oregon Mining has not been available for nearly seventy five years. Included are extremely rare insights into the history and day to day operations of the dragline and bucketline gold dredges that once worked the placer gold fields of South West and North East Oregon in decades gone by. Also included are details into the areas that were worked by gold dredges in Josephine, Jackson, Baker and Grant counties, as well as the economic factors that impacted this mining method. This volume also offers a unique look into the values of river bottom land in relation to both farming and mining, in how farm lands were mined, re-soiled and reclamated after the dredges worked them. Featured are hard to find maps of the gold dredge fields, as well as rare photographs from a bygone era. 8.5" X 11", 86 ppgs. Retail Price: $8.99

Quick Silver Mining in Oregon - Originally published in 1963, this important publication on Oregon Mining has not been available for over fifty years. This publication includes details into the history and production of Elemental Mercury or Quicksilver in the State of Oregon. 8.5" X 11", 238 ppgs. Retail Price: $15.99

Mines of the Greenhorn Mining District of Grant County Oregon - Originally published in 1948, this important publication on Oregon Mining has not been available for over sixty five years. In this publication are rare insights into the mines of the famous Greenhorn Mining District of Grant County, Oregon, especially the famous Morning Mine. Also included are details on the Tempest, Tiger, Bi-Metallic, Windsor, Psyche, Big Johnny, Snow Creek, Banzette and Paramount Mines, as well as prospects in the vicinities in the famous mining areas of Mormon Basin, Vinegar Basin and Desolation Creek. Included are hard to find mine maps and dozens of rare photographs from the bygone era of Grant County's rich mining history. 8.5" X 11", 72 ppgs. Retail Price: $9.99

Geology of the Wallowa Mountains of Oregon: Part I (Volume 1) - Originally published in 1938, this important publication on Oregon Mining has not been available for nearly seventy five years. Included are details on the geology of this unique portion of North Eastern Oregon. This is the first part of a two book series on the area. Accompanying the text are rare photographs and historic maps. **8.5" X 11", 92 ppgs. Retail Price: $9.99**

Geology of the Wallowa Mountains of Oregon: Part II (Volume 2) - Originally published in 1938, this important publication on Oregon Mining has not been available for nearly seventy five years. Included are details on the geology of this unique portion of North Eastern Oregon. This is the first part of a two book series on the area. Accompanying the text are rare photographs and historic maps. **8.5" X 11", 94 ppgs. Retail Price: $9.99**

Field Identification of Minerals For Oregon Prospectors - Originally published in 1940, this important publication on Oregon Mining has not been available for nearly seventy five years. Included in this volume is an easy system for testing and identifying a wide range of minerals that might be found by prospectors, geologists and rockhounds in the State of Oregon, as well as in other locales. Topics include how to put together your own field testing kit and how to conduct rudimentary tests in the field. This volume is written in a clear and concise way to make it useful even for beginners. **8.5" X 11", 158 ppgs. Retail Price: $14.99**

The Bohemia Mining District of Oregon - Originally published in 1900, this important publication on Oregon Mining has not been available for over a century. Included in this volume are important insights into the famous Bohemia Mining District of Oregon, including the histories and locations of important gold mines in the area such as the Ophir Mine, Clarence, Acturas, Peek-a-boo, White Swan, Combination Mine, the Musick Mine, The California, White Ghost, The Mystery, Wall Street, Vesuvius, Story, Lizzie Bullock, Delta, Elsie Dora, Golden Slipper, Broadway, Champion Mine, Knott, Noonday, Helena, White Wings, Riverside and others. Also included are notes on the nearby Blue River Mining District. **8.5" X 11", 58 ppgs. Retail Price: $9.99**

The Gold Fields of Eastern Oregon - Unavailable since 1900, this publication was originally compiled by the Baker City Chamber of Commerce Offering important insights into the gold mining history of Eastern Oregon, "The Gold Fields of Eastern Oregon" sheds a rare light on many of the gold mines that were operating at the turn of the 19th Century in Baker County and Grant County in North Eastern Oregon. Some of the areas featured include the Cable Cove District, Baisely-Elhorn, Granite, Red Boy, Bonanza, Susanville, Sparta, Virtue, Vaughn, Sumpter, Burnt River, Rye Valley and other mining districts. Included is basic information on not only many gold mines that are well known to those interested in Eastern Oregon mining history, but also many mines and prospects which have been mostly lost to the passage of time. Accompanying are numerous rare photos **8.5" X 11", 78 ppgs. Retail Price: $10.99**

Gold Mining in Eastern Oregon - Originally published in 1938, this important publication on Oregon Mining has not been available for over a century. Included in this volume are important insights into the famous mining districts of Eastern Oregon during the late 1930's. Particular attention is given to those gold mines with milling and concentrating facilities in the Greenhorn, Red Boy, Alamo, Bonanza, Granite, Cable Cove, Cracker Creek, Virtue, Keating, Medical Springs, Sanger, Sparta, Chicken Creek, Mormon Basin, Connor Creek, Cornucopia and the Bull Run Mining Districts. Some of the mines featured include the Ben Harrison, North Pole-Columbia, Highland Maxwell, Baisley-Elkhorn, White Swan, Balm Creek, Twin Baby, Gem of Sparta, New Deal, Gleason, Gifford-Johnson, Cornucopia, Record, Bull Run, Orion and others. Of particular interest are the mill flow sheets and descriptions of milling operations of these mines. **8.5" X 11", 68 ppgs. Retail Price: $8.99**

The Gold Belt of the Blue Mountains of Oregon - Originally published in 1901, this important publication on Oregon Mining has not been available for over a century. Included in this volume are rare insights into the gold deposits of the Blue Mountains of North East Oregon, including the history of their early discovery and early production. Extensive details are offered on this important mining area's mineralogy and economic geology, as well as insights into nearby gold placers, silver deposits and copper deposits. Featured are the Elkhorn and Rock Creek mining districts, the Pocahontas district, Auburn and Minersville districts, Sumpter and Cracker Creek, Cable Cove, the Camp Carson district, Granite, Alamo, Greenhorn, Robinsonville, the Upper Burnt River Valley and Bonanza districts, Susanville, Quartzburg, Canyon Creek, Virtue, the Copper Butte district, the North Powder River, Sparta, Eagle Creek, Cornucopia, Pine Creek, Lower Powder River, the Upper Snake River Canyon, Rye Valley, Lower Burnt River Valley, Mormon Basin, the Malheur and Clarks Creek districts, Sutton Creek and others. Of particular interest are important details on numerous gold mines and prospects in these mining districts, including their locations, histories, geology and other important information, as well as information on silver, copper and fire opal deposits. **8.5" X 11", 250 ppgs. Retail Price: $24.99**

<u>Mining in the Cascades Range of Oregon</u> - Originally published in 1938, this important publication on Oregon Mining has not been available for over seventy five years. Included in this volume are rare insights into the gold mines and other types of metal mines in the Cascades Mountain Range of Oregon. Some of the important mining areas covered include the famous Bohemia Mining District, the North Santiam Mining District, Quartzville Mining District, Blue River Mining District, Fall Creek Mining District, Oakridge District, Zinc District, Buzzard-Al Sarena District, Grand Cove, Climax District and Barron Mining District. Of particular interest are important details on over 100 mines and prospects in these mining districts, including their locations, histories, geology and other important information. **8.5″ X 11″, 170 ppgs. Retail Price: $14.99**

<u>Beach Gold Placers of the Oregon Coast</u> - Originally published in 1934, this important publication on Oregon Mining has not been available for over 80 years. Included in this volume are rare insights into the beach gold deposits of the State of Oregon, including their locations, occurance, composition and geology. Of particular interest is information on placer platinum in Oregon's rich beach deposits. Also included are the locations and other information on some famous Oregon beach mines, including the Pioneer, Eagle, Chickamin, Iowa and beach placer mines north of the mouth of the Rogue River. **8.5″ X 11″, 60 ppgs. Retail Price: $8.99**

Idaho Mining Books

Gold in Idaho - Unavailable since the 1940's, this publication was originally compiled by the Idaho Bureau of Mines and includes details on gold mining in Idaho. Included is not only raw data on gold production in Idaho, but also valuable insight into where gold may be found in Idaho, as well as practical information on the gold bearing rocks and other geological features that will assist those looking for placer and lode gold in the State of Idaho. This volume also includes thirteen gold maps that greatly enhance the practical usability of the information contained in this small book detailing where to find gold in Idaho. **8.5″ X 11″, 72 ppgs. Retail Price: $9.99**

Geology of the Couer D'Alene Mining District of Idaho - Unavailable since 1961, this publication was originally compiled by the Idaho Bureau of Mines and Geology and includes details on the mining of gold, silver and other minerals in the famous Coeur D'Alene Mining District in Northern Idaho. Included are details on the early history of the Coeur D'Alene Mining District, local tectonic settings, ore deposit features, information on the mineral belts of the Osburn Fault, as well as detailed information on the famous Bunker Hill Mine, the Dayrock Mine, Galena Mine, Lucky Friday Mine and the infamous Sunshine Mine. This volume also includes sixteen hard to find maps. **8.5″ X 11″, 70 ppgs. Retail Price: $9.99**

The Gold Camps and Silver Cities of Idaho - Originally published in 1963, this important publication on Idaho Mining has not been available for nearly fifty years. Included are rare insights into the history of Idaho's Gold Rush, as well as the mad craze for silver in the Idaho Panhandle. Documented in fine detail are the early mining excitements at Boise Basin, at South Boise, in the Owyhees, at Deadwood, Long Valley, Stanley Basin and Robinson Bar, at Atlanta, on the famous Boise River, Volcano, Little Smokey, Banner, Boise Ridge, Hailey, Leesburg, Lemhi, Pearl, at South Mountain, Shoup and Ulysses, Yellow Jacket and Loon Creek. The story follows with the appearance of Chinese miners at the new mining camps on the Snake River, Black Pine, Yankee Fork, Bay Horse, Clayton, Heath, Seven Devils, Gibbonsville, Vienna and Sawtooth City. Also included are special sections on the Idaho Lead and Silver mines of the late 1800's, as well as the mining discoveries of the early 1900's that paved the way for Idaho's modern mining and mineral industry. Lavishly illustrated with rare historic photos, this volume provides a one of a kind documentary into Idaho's mining history that is sure to be enjoyed by not only modern miners and prospectors who still scour the hills in search of nature's treasures, but also those enjoy history and tromping through overgrown ghost towns and long abandoned mining camps. **8.5″ X 11″, 186 ppgs. Retail Price: $14.99**

Ore Deposits and Mining in North Western Custer County Idaho - Unavailable since 1913, this important publication was originally published by the Us Department of the Interior and has been unavailable for a century. Included are fine details on the geology, geography, gold placers and gold and silver bearing quartz veins of the mining region of North West Custer County, Idaho. Of particular interest is a rare look at the mines and prospects of the region, including those such as the Ramshorn Mine, SkyLark, Riverview, Excelsior, Beardsley, Pacific, Hoosier, Silver Brick, Forest Rose and dozens of others in the Bay Horse Mining District. Also covered are the mines of the Yankee Fork District such as the Lucky Boy, Badger, Black, Enterprise, Charles Dickens, Morrison, Golden Sunbeam, Montana, Golden Gate and others, as well as those in the Loon Mining District. **8.5″ X 11″, 126 ppgs. Retail Price: $12.99**

Gold Rush To Idaho - Unavailable since 1963, this important publication was originally published by the Idaho Bureau of Mines and has been unavailable for 50 years. "Gold Rush To Idaho" revisits the earliest years of the discovery of gold in Idaho Territory and introduces us to the conditions that the pioneer gold seekers met when they blazed a trail through the wilderness of Idaho's mountains and discovered the precious yellow metal at Oro Fino and Pierce. Subsequent rushes followed at places like Elk City, Newsome, Clearwater Station, Florence, Warrens and elsewhere. Of particular interest is a rare look at the hardships that the first miners in Idaho met with during their day to day existences and their attempts to bring law and order to their mining camps. 8.5" X 11", 88 ppgs. Retail Price: $9.99

The Geology and Mines of Northern Idaho and North Western Montana - Unavailable since 1909, this important publication was originally published by the Us Department of the Interior and has been unavailable for a century. Included are fine details on the geology and geography of the mining regions of Northern Idaho and North Western Montana. Of particular interest is a rare look at the mines and prospects of the region, including those in the Pine Creek Mining District, Lake Pend Oreille district, Troy Mining District, Sylvanite District, Cabinet Mining District, Prospect Mining District and the Missoula Valley. Some of the mines featured include the Iron Mountain, Silver Butte, Snowshoe, Grouse Mountain Mine and others. 8.5" X 11", 142 ppgs. Retail Price: $12.99

Mining in the Alturas Quadrangle of Blaine County Idaho - Unavailable since 1922, this important publication was originally published by the Idaho Bureau of Mines and has been unavailable for ninety years. Topics include the geology, rock formations and the formation of ore deposits in this important mining area of Idaho. Of particular focus is information on the local geology, quartz veins and ore deposits of this portion of Idaho. Included are hard to find details, including the descriptions and locations of numerous gold and silver mines in the area including the Silver King, Pilgrim, Columbia, Lone Jack, Sunbeam, Pride of the West, Lucky Boy, Scotia, Atlanta, Beaver-Bidwell and others mines and prospects. 8.5" X 11", 56 ppgs. Retail Price: $8.99

Mining in Lemhi County Idaho - Originally published in 1913, this important book on Idaho Mining has not been available to miners for over a century. Included are rare insights into hundreds of gold, silver, copper and other mines in this famous Idaho mining area. Details include the locations, geology, history, production and other facts of the mines of this region, not only gold and silver hardrock mines, but also gold placer mines, lead-silver deposits, copper mines, cobalt-nickel deposits, tungsten and tin mines . It is lavishly illustrated with hard to find photos of the period and rare mining maps. Some of the vicinities featured include the Nicholia Mining District, Spring Mountain District, Texas District, Blue Wing District, Junction District, McDevitt District, Pratt Creek, Eldorado District, Kirtley Creek, Carmen Creek, Gibbonsville, Indian Creek, Mineral Hill District, Mackinaw, Eureka District, Blackbird District, YellowJacket District, Gravel Range District, Junction District, Parker Mountain and other mining districts. 8.5" X 11", 226 ppgs. Retail Price: $19.99

Utah Mining Books

Fluorite in Utah - Unavailable since 1954, this publication was originally compiled by the USGS, State of Utah and U.S. Atomic Energy Commission and details the mining of fluorspar, also known as fluorite in the State of Utah. Included are details on the geology and history of fluorspar (fluorite) mining in Utah, including details on where this unique gem mineral may be found in the State of Utah. 8.5" X 11", 60 ppgs. Retail Price: $8.99

California Mining Books

The Tertiary Gravels of the Sierra Nevada of California - Mining historian Kerby Jackson introduces us to a classic mining work by Waldemar Lindgren in this important re-issue of The Tertiary Gravels of the Sierra Nevada of California. Unavailable since 1911, this publication includes details on the gold bearing ancient river channels of the famous Sierra Nevada region of California. 8.5" X 11", 282 ppgs. Retail Price: $19.99

The Mother Lode Mining Region of California - Unavailable since 1900, this publication includes details on the gold mines of California's famous Mother Lode gold mining area. Included are details on the geology, history and important gold mines of the region, as well as insights into historic mining methods, mine timbering, mining machinery, mining bell signals and other details on how these mines operated. Also included are insights into the gold mines of the California Mother Lode that were in operation during the first sixty years of California's mining history. 8.5" X 11", 176 ppgs. Retail Price: $14.99

Lode Gold of the Klamath Mountains of Northern California and South West Oregon - Unavailable since 1971, this publication was originally compiled by Preston E. Hotz and includes details on the lode mining districts of Oregon and California's Klamath Mountains. Included are details on the geology, history and important lode mines of the French Gulch, Deadwood, Whiskeytown, Shasta, Redding, Muletown, South Fork, Old Diggings, Dog Creek (Delta), Bully Choop (Indian Creek), Harrison Gulch, Hayfork, Minersville, Trinity Center, Canyon Creek, East Fork, New River, Denny, Liberty (Black Bear), Cecilville, Callahan, Yreka, Fort Jones and Happy Camp mining districts in California, as well as the Ashland, Rogue River, Applegate, Illinois River, Takilma, Greenback, Galice, Silver Peak, Myrtle Creek and Mule Creek districts of South Western Oregon. Also included are insights into the mineralization and other characteristics of this important mining region. 8.5" X 11", 100 ppgs. Retail Price: $10.99

Mines and Mineral Resources of Shasta County, Siskiyou County, Trinity County: California - Unavailable since 1915, this publication was originally compiled by the California State Mining Bureau and includes details on the gold mines of this area of Northern California. Also included are insights into the mineralization and other characteristics of this important mining region, as well as the location of historic gold mines. **8.5" X 11", 204 ppgs. Retail Price: $19.99**

Geology of the Yreka Quadrangle, Siskiyou County, California - Unavailable since 1977, this publication was originally compiled by Preston E. Hotz and includes details on the geology of the Yreka Quadrangle of Siskiyou County, California. Also included are insights into the mineralization and other characteristics of this important mining region. **8.5" X 11", 78 ppgs. Retail Price: $7.99**

Mines of San Diego and Imperial Counties, California - Originally published in 1914, this important publication on California Mining has not been available for a century. This publication includes important information on the early gold mines of San Diego and Imperial County, which were some of the first gold fields mined in California by early Spanish and Mexican miners before the 49ers came on the scene. Included are not only details on early mining methods in the area, production statistics and geological information, but also the location of the early gold mines that helped make California "The Golden State". Also included are details on the mining of other minerals such as silver, lead, zinc, manganese, tungsten, vanadium, asbestos, barite, borax, cement, clay, dolomite, fluospar, gem stones, graphite, marble, salines, petroleum, stronium, talc and others. **8.5" X 11", 116 ppgs. Retail Price: $12.99**

Mines of Sierra County, California - Unavailable since 1920, this publication was originally compiled by the California State Mining Bureau and includes details on the gold mines of Sierra County, California. Also included are insights into the mineralization and other characteristics of this important mining region, as well as the location of historic gold mines. **8.5" X 11", 156 ppgs. Retail Price: $19.99**

Mines of Plumas County, California - Unavailable since 1918, this publication was originally compiled by the California State Mining Bureau and includes details on the gold mines of Plumas County, California. Also included are insights into the mineralization and other characteristics of this important mining region, as well as the location of historic gold mines. **8.5" X 11", 200 ppgs. Retail Price: $19.99**

Mines of El Dorado, Placer, Sacramento and Yuba Counties, California - Originally published in 1917, this important publication on California Mining has not been available for nearly a century. This publication includes important information on the early gold mines of El Dorado County, Placer County, Sacramento County and Yuba County, which were some of the first gold fields mined by the Forty-Niners during the California Gold Rush. Included are not only details on early mining methods in the area, production statistics and geological information, but also the location of the early gold mines that helped make California "The Golden State". Also included are insights into the early mining of chrome, copper and other minerals in this important mining area. **8.5" X 11", 204 ppgs. Retail Price: $19.99**

Mines of Los Angeles, Orange and Riverside Counties, California - Originally published in 1917, this important publication on California Mining has not been available for nearly a century. This publication includes important information on the early gold mines of Los Angeles County, Orange County and Riverside County, which were some of the first gold fields mined in California by early Spanish and Mexican miners before the 49ers came on the scene. Included are not only details on early mining methods in the area, production statistics and geological information, but also the location of the early gold mines that helped make California "The Golden State". **8.5" X 11", 146 ppgs. Retail Price: $12.99**

Mines of San Bernadino and Tulare Counties, California - Originally published in 1917, this important publication on California Mining has not been available for nearly a century. This publication includes important information on the early gold mines of San Bernadino and Tulare County, which were some of the first gold fields mined in California by early Spanish and Mexican miners before the 49ers came on the scene. Included are not only details on early mining methods in the area, production statistics and geological information, but also the location of the early gold mines that helped make California "The Golden State". Also included are details on the mining of other minerals such as copper, iron, lead, zinc, manganese, tungsten, vanadium, asbestos, barite, borax, cement, clay, dolomite, fluospar, gem stones, graphite, marble, salines, petroleum, stronium, talc and others. **8.5" X 11", 200 ppgs. Retail Price: $19.99**

Chromite Mining in The Klamath Mountains of California and Oregon - Unavailable since 1919, this publication was originally compiled by J.S. Diller of the United States Department of Geological Survey and includes details on the chromite mines of this area of Northern California and Southern Oregon. Also included are insights into the mineralization and other characteristics of this important mining region, as well as the location of historic mines. Also included are insights into chromite mining in Eastern Oregon and Montana. **8.5" X 11", 98 ppgs. Retail Price: $9.99**

Mines and Mining in Amador, Calaveras and Tuolumne Counties, California - Unavailable since 1915, this publication was originally compiled by William Tucker and includes details on the mines and mineral resources of this important California mining area. Included are details on the geology, history and important gold mines of the region, as well as insights into other local mineral resources such as asbestos, clay, copper, talc, limestone and others. Also included are insights into the mineralization and other characteristics of this important portion of California's Mother Lode mining region. 8.5" X 11", 198 ppgs. **Retail Price: $14.99**

The Cerro Gordo Mining District of Inyo County California - Unavailable since 1963, this publication was originally compiled by the United States Department of Interior. Included are insights into the mineralization and other characteristics of this important mining region of Southern California. Topics include the mining of gold and silver in this important mining district in Inyo County, California, including details on the history, production and locations of the Cerro Gordo Mine, the Morning Star Mine, Estelle Tunnel, Charles Lease Tunnel, Ignacio, Hart, Crosscut Tunnel, Sunset, Upper Newtown, Newtown, Ella, Perseverance, Newsboy, Belmont and other silver and gold mines in the Cerro Gordo Mining District. This volume also includes important insights into the fossil record, geologic formations, faults and other aspects of economic geology in this California mining district. 8.5" X 11", **104 ppgs. Retail Price: $10.99**

Mining in Butte, Lassen, Modoc, Sutter and Tehama Counties of California - Unavailable since 1917, this publication was originally compiled by the United States Department of Interior. Included are insights into the mineralization and other characteristics of this important mining region of California. Topics include the mining of asbestos, chromite, gold, diamonds and manganese in Butte County, the mining of gold and copper in the Hayden Hill and Diamond Mountain mining districts of Lassen County, the mining of coal, salt, copper and gold in the High Grade and Winters mining districts of Modoc County, gold mining in Sutter County and the mining of gold, chromite, manganese and copper in Tehama County. This volume also includes the production records and locations of numerous mines in this important mining region. 8.5" X 11", 114 ppgs. **Retail Price: $11.99**

Mines of Trinity County California - Originally published in 1965, this important publication on California Mining has not been available for nearly fifty years. This publication includes important information on mines and mining in Trinity County, California, as well insights into the mineralization and geology of this important mining area in Northern California. Included are extensive details on hardrock and placer gold mines and prospects, including charts showing the locations of these historic mines.. 8.5" X 11", 144 ppgs. **Retail Price: $12.99**

Mines of Kern County California - Originally published in 1962, this important publication on California Mining has not been available for nearly fifty years. This publication includes important information on mines and mining in Kern County, California, as well insights into the mineralization and geology of this important mining area in California. Included are extensive details on hardrock and placer gold mines and prospects, including charts showing the locations of these historic mines. 8.5" X 11", 398 ppgs. **Retail Price: $24.99**

Mines of Calaveras County California - Originally published in 1962, this important publication on California Mining has not been available for nearly fifty years. This publication includes important information on mines and mining in Calaveras County, California, as well insights into the mineralization and geology of this important mining area in Northern California. Included are extensive details on hardrock and placer gold mines and prospects, including charts showing the locations of these historic mines. 8.5" X 11", 236 ppgs. **Retail Price: $19.99**

Lode Gold Mining in Grass Valley California - Unavailable since 1940, this publication was originally compiled by the United States Department of Interior. Included are insights into the gold mineralization and other characteristics of this important mining region of Nevada County, California. This volume also includes important insights into the geologic formations, faults and other aspects of economic geology in this California mining district. Of particular interest are the fine details on many hardrock gold mines in the area, including their locations, histories, development and mineralization. Some of the mines featured include the Gold Hill Mine, Massachusetts Hill, Boundary, Peabody, Golden Center, North Star, Omaha, Lone Jack, Homeward Bound, Hartery, Wisconsin, Allison Ranch, Phoenix, Kate Hayes, W.Y.O.D., Empire, Rich Hill, Daisy Hill, Orleans, Sultana, Centennial, Conlin, Ben Franklin, Crown Point and many others. 8.5" X 11", 148 ppgs. **Retail Price: $12.99**

Lode Mining in the Alleghany District of Sierra County California - Unavailable since 1913, this publication was originally compiled by the United States Department of Interior. Included are insights into the mineralization and other characteristics of this important mining region of Sierra County. Included are details on the history, production and locations of numerous hardrock gold mines in this famous California area, including the Tightner Mine, Minnie D., Osceola, Eldorado, Twenty One, Sherman, Kenton, Oriental, Rainbow, Plumbago, Irelan, Gold Canyon, North Fork, Federal, Kate Hardy and others. This volume also includes important insights into the fossil record, geologic formations, faults and other aspects of economic geology in this California mining district. 8.5" X 11", 48 ppgs. **Retail Price: $7.99**

Six Months In The Gold Mines During The California Gold Rush - Unavailable since 1850, this important work is a first hand account of one "49'ers" personal experience during the great California Gold Rush, shedding important light on one of the most exciting periods in the history of not only California, but also the world. Compiled from journals written between 1847 and 1849 by E. Gould Buffum, a native of New York, "Six Months In The Gold Mines During The California Gold Rush" offers a rare look into the day to day lives of the people who came to California to work in her gold mines when the state was still a great frontier. **8.5" X 11", 290 ppgs. Retail Price: $19.99**

Quartz Mines of the Grass Valley Mining District of California - Unavailable since 1867, this important publication has not been available since those days. This rare publication offers a short dissertation on the early hardrock mines in this important mining district in the California Mother Lode region between the 1850's and 1860's. Also included are hard to find details on the mineralization and locations of these mines, as well as how they were operated in those day. **8.5" X 11", 44 ppgs. Retail Price: $8.99**

Alaska Mining Books

Ore Deposits of the Willow Creek Mining District, Alaska - Unavailable since 1954, this hard to find publication includes valuable insights into the Willow Creek Mining District near Hatcher Pass in Alaska. The publication includes insights into the history, geology and locations of the well known mines in the area, including the Gold Cord, Independence, Fern, Mabel, Lonesome, Snowbird, Schroff-O'Neil, High Grade, Marion Twin, Thorpe, Webfoot, Kelly-Willow, Lane, Holland and others. **8.5" X 11", 96 ppgs. Retail Price: $9.99**

The Juneau Gold Belt of Alaska - Unavailable since 1906, this hard to find publication includes valuable insights into the gold mines around Juneau, Alaska. The publication includes important details into the history, geology and locations of the well known gold mines and prospects in the area, including those around Windham Bay, Holkham Bay, Port Snettisham, on Grindstone and Rhine Creeks, Gold Creek, Douglas Island, Salmon Creek, Lemon Creek, Nugget Creek, from the Mendenhall River to Berners Bay, McGinnis Creek, Montana Creek, Peterson Creek, Windfall Creek, the Eagle River, Yankee Basin, Yankee Curve, Kowee Creek and elsewhere. Not only are gold placer mines included, but also hardrock gold mines. **8.5" X 11", 224 ppgs. Retail Price: $19.99**

Arizona Mining Books

Mines and Mining in Northern Yuma County Arizona - Originally published in 1911, this important publication on Arizona Mining has not been available for over a hundred years. Included are rare insights into the gold, silver, copper and quicksilver mines of Yuma County, Arizona together with hard to find maps and photographs. Some of the mines and mining districts featured include the Planet Copper Mine, Mineral Hill, the Clara Consolidated Mine, Viati Mine, Copper Basin prospect, Bowman Mine, Quartz King, Billy Mack, Carnation, the Wardwell and Osbourne, Valensuella Copper, the Mariquita, Colonial Mine, the French American, the New York-Plomosa, Guadalupe, Lead Camp, Mudersbach Copper Camp, Yellow Bird, the Arizona Northern (Salome Strike), Bonanza (Harqua Hala), Golden Eagle, Hercules, Socorro and others. **8.5" X 11", 144 ppgs. Retail Price: $11.99**

The Aravaipa and Stanley Mining Districts of Graham County Arizona - Originally published in 1925, this important publication on Arizona Mining has not been available for nearly ninety years. Included are rare insights into the gold and silver mines of these two important mining districts, together with hard to find maps. **8.5" X 11", 140 ppgs. Retail Price: $11.99**

Gold in the Gold Basin and Lost Basin Mining Districts of Mohave County, Arizona - This volume contains rare insights into the geology and gold mineralization of the Gold Basin and Lost Basin Mining Districts of Mohave County, Arizona that will be of benefit to miners and prospectors. Also included is a significant body of information on the gold mines and prospects of this portion of Arizona. This volume is lavishly illustrated with rare photos and mining maps. **8.5" X 11", 188 ppgs. Retail Price: $19.99**

Mines of the Jerome and Bradshaw Mountains of Arizona - This important publication on Arizona Mining has not been available for ninety years. This volume contains rare insights into the geology and ore deposits of the Jerome and Bradshaw Mountains of Arizona that will be of benefit to miners and prospectors who work those areas. Included is a significant body of information on the mines and prospects of the Verde, Black Hills, Cherry Creek, Prescott, Walker, Groom Creek, Hassayampa, Bigbug, Turkey Creek, Agua Fria, Black Canyon, Peck, Tiger, Pine Grove, Bradshaw, Tintop, Humbug and Castle Creek Mining Districts. This volume is lavishly illustrated with rare photos and mining maps. **8.5" X 11", 218 ppgs. Retail Price: $19.99**

The Ajo Mining District of Pima County Arizona - This important publication on Arizona Mining has not been available for nearly seventy years. This volume contains rare insights into the geology and mineralization of the Ajo Mining District in Pima County, Arizona and in particular the famous New Cornelia Mine. **8.5" X 11", 126 ppgs. Retail Price: $11.99**

Mining in the Santa Rita and Patagonia Mountains of Arizona - Originally published in 1915, this important publication on Arizona Mining has not been available for nearly a century. Included are rare insights into hundreds of gold, silver, copper and other mines in this famous Arizona mining area. Details include the locations, geology, history, production and other facts of the mines of this region. **8.5" X 11", 394 ppgs. Retail Price: $24.99**

Mining in the Bisbee Quadrangle of Arizona - Originally published in 1906, this important publication on Arizona Mining has not been available for nearly a century. Included are rare insights into hundreds of gold, silver, copper and other mines in this famous Arizona mining area. Details include the locations, geology, history, production and other facts of the mines of this important mining region. **8.5" X 11", 188 ppgs. Retail Price: $14.99**

Montana Mining Books

A History of Butte Montana: The World's Greatest Mining Camp - First published in 1900 by H.C. Freeman, this important publication sheds a bright light on one of the most important mining areas in the history of The West. Together with his insights, as well as rare photographs of the periods, Harry Freeman describes Butte and its vicinity from its early beginnings, right up to its flush years when copper flowed from its mines like a river. At the time of publication, Butte, Montana was known worldwide as "The Richest Mining Spot On Earth" and produced not only vast amounts of copper, but also silver, gold and other metals from its mines. Freeman illustrates, with great detail, the most important mines in the vicinity of Butte, providing rare details on their owners, their history and most importantly, how the mines operated and how their treasures were extracted. Of particular interest are the dozens of rare photographs that depict mines such as the famous Anaconda, the Silver Bow, the Smoke House, Moose, Paulin, Buffalo, Little Minah, the Mountain Consolidated, West Greyrock, Cora, the Green Mountain, Diamond, Bell, Parnell, the Neversweat, Nipper, Original and many others. **8.5" X 11", 142 ppgs. Retail Price: $12.99**

The Butte Mining District of Montana - This important publication on Montana Mining has not been available for over a century. Included are rare insights into the gold, copper and silver mines of Butte, Montana together with hard to find maps and photographs. Some of the topics include the early history of gold, silver and copper mining in the Butte area, insight into the geology of its mining areas, the local distribution of gold, silver and copper ores, as well their composition and how to identify them. Also included are detailed facts about the mines in the Butte Mining District, including the famous Anaconda Mine, Gagnon, Parrot, Blue Vein, Moscow, Poulin, Stella, Buffalo, Green Mountain, Wake Up Jim, the Diamond-Bell Group, Mountain Consolidated, East Greyrock, West Greyrock, Snowball, Corra, Speculator, Adirondack, Miners Union, the Jessie-Edith May Group, Otisco, Iduna, Colorado, Lizzie, Cambers, Anderson, Hesperus, Preferencia and dozens of others. **8.5" X 11", 298 ppgs. Retail Price: $24.99**

Mines of the Helena Mining Region of Montana - This important publication on Montana Mining has not been available for over a century. Included are rare insights into the gold, copper and silver mines of the vicinity of Helena, Montana, including the Marysville Mining District, Elliston Mining District, Rimini Mining District, Helena Mining District, Clancy Mining District, Wickes Mining District, Boulder and Basin Mining Districts and the Elkhorn Mining District. Some of the topics include the early history of gold, silver and copper mining in the Helena area, insight into the geology of its mining areas, the local distribution of gold, silver and copper ores, as well their composition and how to identify them. Also included are detailed facts, history, geology and locations of over one hundred gold, silver and copper mines in the area . **8.5" X 11", 162 ppgs, Retail Price: $14.99**

Mines and Geology of the Garnet Range of Montana - This important publication on Montana Mining has not been available for over a century. Included are rare insights into the gold, copper and silver mines of the vicinity of this important mining area of Montana. Some of the topics include the early history of gold, silver and copper mining in the Garnet Mountains, insight into the geology of its mining areas, the local distribution of gold, silver and copper ores, as well their composition and how to identify them. Also included are detailed facts, history, geology and locations of numerous gold, silver and copper mines in the area . **8.5" X 11", 100 ppgs, Retail Price: $11.99**

Mines and Geology of the Philipsburg Quadrangle of Montana - This important publication on Montana Mining has not been available for over a century. Included are rare insights into the gold, copper and silver mines of the vicinity of this important mining area of Montana. Some of the topics include the early history of gold, silver and copper mining in the Philipsburg Quadrangle, insight into the geology of its mining areas, the local distribution of gold, silver and copper ores, as well their composition and how to identify them. Also included are detailed facts, history, geology and locations of over one hundred gold, silver and copper mines in the area **8.5" X 11", 290 ppgs, Retail Price: $24.99**

Geology of the Marysville Mining District of Montana - Included are rare insights into the mining geology of the Marysville Mining District. Some of the topics include the early history of gold, silver and copper mining in the area, insight into the geology of its mining areas, the local distribution of gold, silver and copper ores, as well their composition and how to identify them. Also included are detailed facts, history, geology and locations of gold, silver and copper mines in the area **8.5" X 11", 198 ppgs, Retail Price: $19.99**

The Geology and Mines of Northern Idaho and North Western Montana

See listing under Idaho.

Nevada Mining Books

The Bull Frog Mining District of Nevada - Unavailable since 1910, this publication was originally compiled by the United States Department of Interior. This volume also includes important insights into the geologic formations, faults and other aspects of economic geology in this Nevada mining district. Of particular interest are the fine details on many mines in the area, including their locations, histories, development and mineralization. Some of the mines featured include the National Bank Mine, Providence, Gibraltor, Tramps, Denver, Original Bullfrog, Gold Bar, Mayflower, Homestake-King and other mines and prospects. **8.5″ X 11″, 152 ppgs, Retail Price: $14.99**

History of the Comstock Lode - Unavailable since 1876, this publication was originally released by John Wiley & Sons. This volume also includes important insights into the famous Comstock Lode of Nevada that represented the first major silver discovery in the United States. During its spectacular run, the Comstock produced over 192 million ounces of silver and 8.2 million ounces of gold. Not only did the Comstock result in one of the largest mining rushes in history and yield immense fortunes for its owners, but it made important contributions to the development of the State of Nevada, as well as neighboring California. Included here are important details on not only the early development and history of the Comstock, but also rare early insight into its mines, ore and its geology.**8.5″ X 11″, 244 ppgs, Retail Price: $19.99**

Colorado Mining Books

Ores of The Leadville Mining District - Unavailable since 1926, this publication was originally compiled by the United States Department of Interior. This volume also includes important insights into the ores and mineralization of the Leadville Mining District in Colorado. Topics include historic ore prospecting methods, local geology, insights into ore veins and stockworks, the local trend and distribution of ore channels, reverse faults, shattered rock above replacement ore bodies, mineral enrichment in oxidized and sulphide zones and more. **8.5″ X 11″, 66 ppgs, Retail Price: $8.99**

Mining in Colorado - Unavailable since 1926, this publication was originally compiled by the United States Department of Interior. This volume also includes important insights into the mining history of Colorado from its early beginnings in the 1850's right up to the mid 1920's. Not only is Colorado's gold mining heritage included, but also its silver, copper, lead and zinc mining industry. Each mining area is treated separately, detailing the development of Colorado's mines on a county by county basis. **8.5″ X 11″, 284 ppgs, Retail Price: $19.99**

Gold Mining in Gilpin County Colorado - Unavailable since 1876, this publication was originally compiled by the Register Steam Printing House of Central City, Colorado. A rare glimpse at the gold mining history and early mines of Gilpin County, Colorado from their first discovery in the 1850's up to the "flush years" of the mid 1870's. Of particular interest is the history of the discovery of gold in Gilpin County and details about the men who made those first strikes. Special focus is given to the early gold mines and first mining districts of the area, many of which are not detailed in other books on Colorado's gold mining history. **8.5″ X 11″, 156 ppgs, Retail Price: $12.99**

Mining in the Gold Brick Mining District of Colorado - Important insights into the history of the Gold Brick Mining District, as well as its local geography and economic geology. Also included are the histories and locations of historic mines in this important Colorado Mining District, including the Cortland, Carter, Raymond, Gold Links, Sacramento, Bassick, Sandy Hook, Chronicle, Grand Prize, Chloride, Granite Mountain, Lucille, Gray Mountain, Hilltop, Maggie Mitchell, Silver Islet, Revenue, Roosevelt, Carbonate King and others. In addition to hardrock mining, are also included are details on gold placer mining in this portion of Colorado. **8.5″ X 11″, 140 ppgs, Retail Price: $12.99**

Washington Mining Books

The Republic Mining District of Washington - Unavailable since 1910, this important publication was originally published by the Washington Geologic Survey and has been unavailable for a century. Topics include the geology, rock formations and the formation of ore deposits in this important mining area of Washington State. Also included are hard to find details on the geology, history and locations of dozens of mines in the area. Some of the mines featured include the New Republic Mine, Ben Hur, Morning Glory, the South Republic Mine, Quilp, Surprise, Black Tail, Lone Pine, San Poil, Mountain Lion, Tom Thumb, Elcaliph and many others. **8.5″ X 11″, 94 ppgs, Retail Price: $10.99**

Wyoming Mining Books

Mining in the Laramie Basin of Wyoming - Unavailable since 1909, this publication was originally compiled by the United States Department of Interior. Also included are insights into the mineralization and other characteristics of this important mining region, especially in regards to coal, limestone, gypsum, bentonite clay, cement, sand, clay and copper. 8.5" X 11", 104 ppgs, Retail Price: $11.99

New Mexico Mining Books

The Mogollon Mining District of New Mexico - Unavailable since 1927, this important publication was originally published by the US Department of Interior and has been unavailable for 80 years. Topics include the geology, rock formations and the formation of ore deposits in this important mining area in New Mexico. Of particular focus is information on the history and production of the ore deposits in this area, their form and structure, vein filling, their paragenesis, origins and ore shoots, as well as oxidation and supergene enrichment. Also included are hard to find details, including the descriptions and locations of numerous gold, silver and other types of mines, including the Eureka, Pacific, South Alpine, Great Western, Enterprise, Buffalo, Mountain View, Floride, Gold Dust, Last Chance, Deadwood, Confidence, Maud S., Deep Down, Little Fanney, Trilby, Johnson, Alberta, Comet, Golden Eagle, Cooney, Queen, the Iron Crown, Eberle, Clifton, Andrew Jackson mine, Mascot and others. 8.5" X 11", 144 ppgs, Retail Price: $12.99

The Percha Mining District of Kingston New Mexico - Unavailable since 1883, this important publication was originally published by the Kingston Tribune and has been unavailable for over one hundred and thirty five years. Having been written during the earliest years of gold and silver mining in the Percha Mining District, unlike other books on the subject, this work offers the unique perspective of having actually been written while the early mining history of this area was still being made. In fact, the work was written so early in the development of this area that many of the notable mines in the Percha District were less than a few years old and were still being operated by their original discoverers with the same enthusiasm as when they were first located. Included are hard to find details on the very earliest gold and silver mines of this important mining district near Kingston in Sierra County, New Mexico. 8.5" X 11", 68 ppgs, Retail Price: $9.99

East Coast Mining Books

The Gold Fields of the Southern Appalachians - Unavailable since 1895, this important publication was originally published by the US Department of Interior and has been unavailable for nearly 120 years. Topics include the geology, rock formations and the formation of ore deposits in this important mining area of the American South. Of particular focus is information on the history and statistics of the ore deposits in this area, their form and structure and veins. Also included are details on the placer gold deposits of the region. The gold fields of the Georgian Belt, Carolinian Belt and the South Mountain Mining District of North Carolina are all treated in descriptive detail. Included are hard to find details, including the descriptions and locations of numerous gold mines in Georgia, North Carolina and elsewhere in the American South. Also included are details on the gold belts of the British Maritime Provinces and the Green Mountains. 8.5" X 11", 104 ppgs, Retail Price: $9.99

Gold Rush Tales Series

Millions in Siskiyou County Gold - In this first volume of the "Gold Rush Tales" series, leading mining historian and editor Kerby Jackson, introduces us to the story of how millions of dollars worth of gold was discovered in Siskiyou County during the California Gold Rush. Lavishly illustrated with photos from the 19th Century, this hard to find information was first published in 1897 and sheds important light onto the gold rush era in Siskiyou County, California and the experiences of the men who dug for the gold and actually found it. 8.5" X 11", 82 ppgs, Retail Price: $9.99

The California Rand in the Days of '49 - In this second volume of the "Gold Rush Tales" series, leading mining historian and editor Kerby Jackson, introduces us to four tales from the California Gold Rush. Lavishly illustrated with photos from the 19th Century, this hard to find information was first published in 1890's and includes the stories of "California's Rand", details about Chinese miners, how one early miner named Baker struck it rich and also the story of Alphonzo Bowers, who invented the first hydraulic gold dredge. 8.5" X 11", 54 ppgs, Retail Price: $9.99

More Mining Books

Prospecting and Developing A Small Mine - Topics covered include the classification of varying ores, how to take a proper ore sample, the proper reduction of ore samples, alluvial sampling, how to understand geology as it is applied to prospecting and mining, prospecting procedures, methods of ore treatment, the application of drilling and blasting in a small mine and other topics that the small scale miner will find of benefit. 8.5" X 11", 112 ppgs, Retail Price: $11.99

Timbering For Small Underground Mines - Topics covered include the selection of caps and posts, the treatment of mine timbers, how to install mine timbers, repairing damaged timbers, use of drift supports, headboards, squeeze sets, ore chute construction, mine cribbing, square set timbering methods, the use of steel and concrete sets and other topics that the small underground miner will find of benefit. This volume also includes twenty eight illustrations depicting the proper construction of mine timbering and support systems that greatly enhance the practical usability of the information contained in this small book. **8.5" X 11", 88 ppgs. Retail Price: $10.99**

Timbering and Mining - A classic mining publication on Hard Rock Mining by W.H. Storms. Unavailable since 1909, this rare publication provides an in depth look at American methods of underground mine timbering and mining methods. Topics include the selection and preservation of mine timbers, drifting and drift sets, driving in running ground, structural steel in mine workings, timbering drifts in gravel mines, timbering methods for driving shafts, positioning drill holes in shafts, timbering stations at shafts, drainage, mining large ore bodies by means of open cuts or by the "Glory Hole" system, stoping out ore in flat or low lying veins, use of the "Caving System", stoping in swelling ground, how to stope out large ore bodies, Square Set timbering on the Comstock and its modifications by California miners, the construction of ore chutes, stoping ore bodies by use of the "Block System", how to work dangerous ground, information on the "Delprat System" of stoping without mine timbers, construction and use of headframes and much more. This volume provides a reference into not only practical methods of mining and timbering that may be employed in narrow vein mining by small miners today, but also rare insights into how mines were being worked at the turn of the 19th Century. **8.5" X 11", 288 ppgs. Retail Price: $24.99**

A Study of Ore Deposits For The Practical Miner - Mining historian Kerby Jackson introduces us to a classic mining publication on ore deposits by J.P. Wallace. First published in 1908, it has been unavailable for over a century. Included are important insights into the properties of minerals and their identification, on the occurrence and origin of gold, on gold alloys, insights into gold bearing sulfides such as pyrites and arscnopyrites, on gold bearing vanadium, gold and silver tellurides, lead and mercury tellurides, on silver ores, platinum and iridium, mercury ores, copper ores, lead ores, zinc ores, iron ores, chromium ores, manganese ores, nickel ores, tin ores, tungsten ores and others. Also included are facts regarding rock forming minerals, their composition and occurrences, on igneous, sedimentary, metamorphic and intrusive rocks, as well as how they are geologically disturbed by dikes, flows and faults, as well as the effects of these geologic actions and why they are important to the miner. Written specifically with the common miner and prospector in mind, the book will help to unlock the earth's hidden wealth for you and is written in a simple and concise language that anyone can understand. **8.5" X 11", 366 ppgs. Retail Price: $24.99**

Mine Drainage - Unavailable since 1896, this rare publication provides an in depth look at American methods of underground mine drainage and mining pump systems. This volume provides a reference into not only practical methods of mining drainage that may be employed in narrow vein mining by small miners today, but also rare insights into how mines were being worked at the turn of the 19th Century. **8.5" X 11", 218 ppgs. Retail Price: $24.99**

Fire Assaying Gold, Silver and Lead Ores - Unavailable since 1907, this important publication was originally published by the Mining and Scientific Press and was designed to introduce miners and prospectors of gold, silver and lead to the art of fire assaying. Topics include the fire assaying of ores and products containing gold, silver and lead; the sampling and preparation of ore for an assay; care of the assay office, assay furnaces; crucibles and scorifiers; assay balances; metallic ores; scorification assays; cupelling; parting' crucible assays, the roasting of ores and more. This classic provides a time honored method of assaying put forward in a clear, concise and easy to understand language that will make it a benefit to even beginners. **8.5" X 11", 96 ppgs. Retail Price: $11.99**

Methods of Mine Timbering - Originally published in 1896, this important publication on mining engineering has not been available for nearly a century. Included are rare insights into historical methods of timbering structural support that were used in underground metal mines during the California that still have a practical application for the small scale hardrock miner of today. **8.5" X 11", 94 ppgs. Retail Price: $10.99**

The Enrichment of Copper Sulfide Ores - First published in 1913, it has been unavailable for over a century. Topics include the definition and types of ore enrichment, the oxidation of copper ores, the precipitation of metallic sulfides. Also included are the results of dozens of lab experiments pertaining to the enrichment of sulfide ores that will be of interest to the practical hard rock mine operator in his efforts to release the metallic bounty from his mine's ore. **8.5" X 11", 92 ppgs. Retail Price: $9.99**

A Study of Magmatic Sulfide Ores - Unavailable since 1914, this rare publication provides an in depth look at magmatic sulfide ores. Some of the topics included are the definition and classification of magmatic ores, descriptions of some magmatic sulfide ore deposits known at the time of publication including copper and nickel bearing pyrrohitic ore bodies, chalcopyrite-bornite deposits, pyritic deposits, magnetite-ileminite deposits, chromite deposits and magmatic iron ore deposits. Also included are details on how to recognize these types of ore deposits while prospecting for valuable hardrock minerals. **8.5" X 11", 138 ppgs. Retail Price: $11.99**

The Cyanide Process of Gold Recovery - Unavailable since 1894 and released under the name "The Cyanide Process: Its Practical Application and Economical Results", this rare publication provides an in depth look at the early use of cyanide leaching for gold recovery from hardrock mine ores. This volume provides a reference into the early development and use of cyanide leaching to recover gold. **8.5" X 11", 162 ppgs. Retail Price: $14.99**

California Gold Milling Practices - Unavailable since 1895 and released under the name "California Gold Practices", this rare publication provides an in depth look at early methods of milling used to reduce gold ores in California during the late 19th century. This volume provides a reference into the early development and use of milling equipment during the earliest years of the California Gold Rush up to the age of the Industrial Revolution. Much of the information still applies today and will be of use to small scale miners engaging in hardrock mining. **8.5" X 11", 104 ppgs. Retail Price: $10.99**

www.ingramcontent.com/pod-product-compliance
Lightning Source LLC
Chambersburg PA
CBHW080253180526
45167CB00006B/2514